T0230454

SpringerBriefs in Energy

SpringerBriefs in Energy presents concise summaries of cutting-edge research and practical applications in all aspects of Energy. Featuring compact volumes of 50 to 125 pages, the series covers a range of content from professional to academic. Typical topics might include:

- A snapshot of a hot or emerging topic
- A contextual literature review
- A timely report of state-of-the art analytical techniques
- An in-depth case study
- A presentation of core concepts that students must understand in order to make independent contributions.

Briefs allow authors to present their ideas and readers to absorb them with minimal time investment.

Briefs will be published as part of Springer's eBook collection, with millions of users worldwide. In addition, Briefs will be available for individual print and electronic purchase. Briefs are characterized by fast, global electronic dissemination, standard publishing contracts, easy-to-use manuscript preparation and formatting guidelines, and expedited production schedules. We aim for publication 8–12 weeks after acceptance.

Both solicited and unsolicited manuscripts are considered for publication in this series. Briefs can also arise from the scale up of a planned chapter. Instead of simply contributing to an edited volume, the author gets an authored book with the space necessary to provide more data, fundamentals and background on the subject, methodology, future outlook, etc.

SpringerBriefs in Energy contains a distinct subseries focusing on Energy Analysis and edited by Charles Hall, State University of New York. Books for this subseries will emphasize quantitative accounting of energy use and availability, including the potential and limitations of new technologies in terms of energy returned on energy invested. The second distinct subseries connected to SpringerBriefs in Energy, entitled Computational Modeling of Energy Systems, is edited by Thomas Nagel, and Haibing Shao, Helmholtz Centre for Environmental Research - UFZ, Leipzig, Germany. This sub-series publishes titles focusing on the role that computer-aided engineering (CAE) plays in advancing various engineering sectors, particularly in the context of transforming energy systems towards renewable sources, decentralized landscapes, and smart grids.

All Springer brief titles should undergo standard single-blind peer-review to ensure high scientific quality by at least two experts in the field.

Federico Moretta · Giulia Bozzano

Mathematical and Statistical Approaches for Anaerobic Digestion Feedstock Optimization

 Springer

Federico Moretta
Department of Chemistry, Materials
and Chemical Engineering
Politecnico di Milano
Milan, Italy

Giulia Bozzano
Department of Chemistry, Materials
and Chemical Engineering
Politecnico di Milano
Milan, Italy

ISSN 2191-5520   ISSN 2191-5539 (electronic)
SpringerBriefs in Energy
ISBN 978-3-031-56459-8   ISBN 978-3-031-56460-4 (eBook)
https://doi.org/10.1007/978-3-031-56460-4

This Springer imprint is published by the registered company Springer Nature Switzerland AG
The registered company address is: Gewerbestrasse 11, 6330 Cham, Switzerland

Paper in this product is recyclable.

Preface

Circular economy and waste valorization together with renewable energy production are a key point for the reduction of human activities' impact on earth and climate equilibrium. Many energy-producing technologies exploit biomasses of different types, preferably coming from residuals and wastes. One of these is anaerobic digestion which produces biogas, a gaseous mixture with a general composition, in the best cases, of 60% methane and 40% biogas. This composition, in facts, is not always reached whatever it is the biomasses blending. Many synergies and antagonisms are occurring in the anaerobic co-digestion of biomasses coming from feedstocks of different natures. Consequently, for an optimized and stable process, it is important to evaluate and monitor the principal parameters that better describe the characteristics and behavior of the biomass itself. This book provides a good overview of the biomass characterization procedure from a mathematical perspective and guides the reader to elaborate a blending optimization routine, to reach a much stable and productive process.

Milan, Italy

Federico Moretta
Giulia Bozzano

Contents

Chapter 1
Substrate General Characteristics

A key point in fermentative processes is the Feedstock composition. It can be divided into two categories: biomass as inoculum or fresh inflow to the process. The inoculum is essential for a stable and productive process, since, if well-tailored, it provides the right amount of microorganisms to run the process [1, 2]. Consequently, setting up a proper biomass is a crucial factor in a productive scenario. It is then necessary to understand and look for its best characteristics for the purpose.

1.1 Macromolecule Content

AD feedstocks can be characterized as mainly composed of three classes of macro-molecules: proteins, carbohydrates, and lipids. Compounds like proteins, fats, sugars, starch, and easily degradable carbohydrates are converted in a relatively short time by microorganisms, due to their high biodegradability; conversely, the presence of high-carbon-content biomasses and lignocellulosic substrates—composed of hardly degradable polysaccharides such as lignin, cellulose, and hemicellulose—make the digestion difficult and reduce the overall biodegradability of the organic matter [3, 4].

Other macronutrients that can affect the process are inorganic compounds such as heavy metals and artificial contaminants, i.e., antibiotics from livestock manures [5]. It is therefore important to modulate these properties and, if it is not possible, to avoid the presence of negative factors, diluting the feedstock as much as possible [6].

F. Moretta and G. Bozzano, *Mathematical and Statistical Approaches for Anaerobic Digestion Feedstock Optimization*, SpringerBriefs in Energy, https://doi.org/10.1007/978-3-031-56460-4_1

1.2   Micronutrients Content

Biological processes are carried out thanks to the presence of specific bacterial strains. To maintain a stable and productive process, it is necessary to feed these microorganisms with the right amount of micronutrients. Solely, these are defined as the raw content of specific chemical species, which differ between strains. Compounds that are generally widely recognized to be biomass nutrients are phosphorous, nitrogen, potassium and carbon [6–8]. Generally, these compounds are used by bacteria through macromolecules degradation (e.g., nitrogen from amino acids), and some of them are in the form of minerals, bounded with magnesium or iron [9].

Feedstocks fed to the reactor usually already contain the bacteria needed. However, different substrates may contain also different quantities and types of microbial species: livestock wastes, for example, usually contain high concentrations of micronutrients, while agro-industrial and food wastes may show a lack in microbial concentrations, altering the strains content [10]. So, tailoring their content by blending complementary feedstocks or using additives can increase the process performances and stabilities.

1.3   Total Solids and Volatile Solids

The Total Solids (TS), expressed as mass fraction of the total substrate weight [%w/w], represent the content of dry matter contained in the feedstock. Experimentally, they are defined as the amount of solids remaining after heating the sample at 105 °C [11]. Depending on the substrate, the quantity of TS that it contains may vary from few point percentages to near 100%, and often substrates need to be diluted to reach the desired solids concentration in the digester [12]. On the other hand, the Volatile Solids (VS), expressed as mass fraction of the total solids content [%w/w$_{TS}$], represents instead the organic fraction of TS. In other terms, what remains after the water content and inorganic matter have been removed from the substrate. The higher is VS content, the higher the organic content of the substrate [13]. Usually, substrates with dry organic content < 60% are rarely considered as valuable for fermentative process due to their poor content of organic matter [5]. However, this parameter doesn't account for the *real* biodegradability of the substrate, neglecting the fact that not all the organic matter can be degraded within the residence time due to the presence of lignocellulosic compounds and other high-carbon-content material.

For example, the higher the TS inside an anaerobic digestion, the higher the methane productivity per unit volume of the reactor (expressed as [$m^3_{CH_4}/m^3_{reactor}/d$]); however, by increasing solids concentration, an over-accumulation of VFAs could be observed, inhibiting bioprocesses as methanogenesis and lowering the efficiency of the process in terms of methane production per mass of VS added (i.e., methane yield, [$m^3_{CH_4}/kg_{VS}/d$]) [14].

In addition, high solid content makes the management of the influent stream more difficult due to the high viscosity, causing poor heat and mass transfer and uneven distribution during stirring, making it necessary to adopt a more complex reactor's design [15]. Therefore, to improve the substrate utilization, at expenses of higher reactor volumes, many research studies have demonstrated that wet digestion (TS < 10%) is the most convenient option, and an optimal TS within this range usually exists [16]. For example, mixtures of municipal solid wastes and sewage sludge in a mass ratio of 60:40 represent an optimal conditions for methane yields were achieved for a concentration of TS of 5% and any further increase would imply a methane yield reduction [14].

1.4 Carbon—Nitrogen Ratio

The ratio between organic carbon and nitrogen in the feedstock (C/N) is a dimensionless parameter that is commonly used to indicate the nutrient content related to the organic matter content. Moreover, it stresses the fact that not only the organic carbon content but also nitrogen content represents a crucial factor in fermentative processes, especially in biogas production [17].

In fact, to obtain good methane yields, it's been demonstrated that C/N should be maintained in a range between about 20 and 40. Below this range, substrates result rich in proteins, that are the main nitrogen source, which by degradation may cause an increase in ammonia concentration that inhibits the digestion process due to an impediment of microbial growth. Above this range, instead, the substrate results rich in carbon sources, leading to the production of high concentrations of VFAs, which are another cause of inhibition due to methanogenic bacteria deactivation [18].

C/N ratio is one of the main parameters affecting the anaerobic digestion process, but often feedstocks are characterized by non-optimal C/N, therefore many studies have been carried out about the maximization of methane yield by co-digesting substrates so that the global carbon to nitrogen ratio falls within the optimal range. For example, it has been investigated the possibility of improving the methane yield of a mixture of dairy manure, chicken manure and wheat straw through the optimization of the C/N ratio [19]. Similarly, [20] performed BMP tests of ternary mixtures of sugar beet root waste, cow dung and poultry manure and concluded that maximum methane yield was obtained with an overall C/N ratio of 26.24.

Like VS, this parameter considers the total amount of organic carbon, of which just a part is biodegradable. To exclude the carbon that is not specifically affected by microorganisms, an available carbon/nitrogen ratio has been proposed in the literature [21].

1.5   pH

pH is another important parameter since it influences the solubilization of organic matter and contributes to the creation of a favorable environment for microbes [18].

The microorganism acting in fermentative processes are characterized by different optimal pH ranges: hydrolytic and acidogenic bacteria prefer a pH within the range 5.5–6.5, while the optimal pH for methanogenic bacteria is near 7 [4, 22–27]. All things considered, many researchers have found out that maintaining a pH between 6.8 and 7.5 is preferable, then the feedstock's pH should be maintained inside this optimal range too. Many research studies that support this statement have been carried out. For example, It has been investigated the effect of initial pH in the digestion of food waste and rice straw, obtaining an optimal methane yield at a C/N ratio of 30 and an initial pH of 7.3 [28].

1.6   Organic Loading Rate and Hydraulic Retention Time

In continuous processes, the organic loading rate (OLR), expressed as [$kg_{VS}/m^3/d$], represents the amount of dry organic solids fed to the reactor per unit volume and unit time. This parameter is strictly related to the hydraulic retention time (HRT), which represents the average time in days that the influent spends inside the reactor, also known as reactor residence time and expressed in [d].

As said, OLR is strictly connected to the HRT and to the TS and VS content inside the reactor: while dealing with a fixed volume continuous reactor, higher OLRs would correspond to shorter HRTs and higher TS content. Similarly, to what happens with high TS content, if the ORL value is too high, overloading is observed leading to an accumulation of VFAs and performances reduction. The OLR, should be maintained within an optimal range [18], and many research studies have been carried out for the determination of optimal OLRs for a variety of substrates and processes [29]. For example, it has been studied [30] that the co-digestion of mixtures of rice straw and cow manure, at a ratio of 50:50 by weight, by varying OLR from 3 to 12 $kg_{VS}/m^3/d$, it has been concluded that the best performances is achieved at 6 $kg_{VS}/m^3/d$, while at higher values the process is inhibited by the accumulation of VFAs. Similarly, it has been found an optimal OLR range of 5–6 $kg_{VS}/m^3/d$ for the digestion of a mixture of rice husk and food waste [31].

At industrial level, however, the inlet flow rate of continuous digesters is solely decided by external factors including pumps' capacity, logistics and strategic issues. Therefore, it is difficult to optimize the process acting only on OLR or HRT values.

1.7 Biodegradability

The biodegradability (BD) is a measure of the fraction of a substrate that is available to be degraded to be effectively converted to further products from microorganism. Indeed, the organic fraction of substrates may be composed both by readily degradable components such as simple carbohydrates, proteins, and lipids, and by hardly degradable fractions, represented by the lignocellulosic components, whose bioavailability is lower and can hardly be degraded. Generally, the higher the biodegradability, the higher and faster the organic matter breakdown.

Many definitions of biodegradability have been used in literature, which are reported below.

Ratio between the Biological Oxygen Demand (BOD) and the Total Chemical Oxygen Demand (COD$_t$) [32]:

$$BD_{BOD} = \frac{BOD}{COD_t} \tag{1.1}$$

The COD$_t$ represents the amount of oxygen present in a sample that can be consumed in a reaction with oxidizing agents, thus it is a measure of the organic fraction of the substrate. The BOD, instead, represents the degradable fraction of the COD$_t$, and thus a measure of the biodegradable organics present in a substrate [17]. If divided by the COD$_t$, therefore, the obtained value represents the biodegradability of the substrate.

Percentage of removed VS [33–35]:

$$BD_{VS} = \frac{VS_{feedstock} - VS_{digestate}}{VS_{feedstock}} \tag{1.2}$$

The biodegradability can be defined as the percentage of VS that has been consumed—i.e., turned into methane—during the digestion process, and can be therefore expressed as the ratio of the difference of VS between initial and final condition, and the VS of the feedstock.

Percentage of removed total COD [33, 34]:

$$BD_{COD} = \frac{COD_{feedstock} - COD_{digestate}}{COD_{feedstock}} \tag{1.3}$$

The biodegradability can also be defined as the COD reduction between the initial and final condition, divided by the initial COD in the feedstock. The meaning of this expression is analogue to the one of BD_{VS}.

Ratio between the Theoretical Biomethane Potential (TBMP) and the Experimental Biomethane Potential (EBMP) [20, 36, 37]:

$$BD_{BMP} = \frac{EBMP}{TBMP} \tag{1.4}$$

Finally, the biodegradability can be defined as the ratio between the cumulative methane yield obtained through a BMP (Biomethane Potential) test and the cumulative yield calculated by theoretical methods. Both TBMP and EBMP are expressed in $[\text{mL}_{CH_4}/g_{VS}]$.

1.8 Theoretical Biomethane Potential

The theoretical biomethane potential (TBMP) represents the theoretical methane yield that could be achieved if the organic matter composing the feedstock would be completely degraded, and if the use of the substrate by microorganisms as an energy source is assumed as insignificant [38]. Several methods can be used to calculate the TBMP, among which two are the most used ones: one based on substrate's elemental composition ($TBMP_{el}$), and one based on organic fractions composition ($TBMP_{org}$).

1.9 TBMP from Elemental Composition

The elemental composition of the biomass is represented by the percentages of carbon, hydrogen, oxygen, nitrogen, sulfur present in a certain substrate.

One of the main formulas used for the estimation of the TBMP is the Buswell equation [39], based on the assumption that the organic matter, expressed with chemical formula $C_n H_a O_b$, is completely degraded to methane and carbon dioxide through an approximate redox reaction involving water, represented by the empirical equation:

$$C_n H_a O_b + \left(n - \frac{a}{4} - \frac{b}{2}\right)H_2O \rightarrow \left(\frac{n}{2} + \frac{a}{8} - \frac{b}{4}\right) \cdot CH_4 + \left(\frac{n}{2} - \frac{a}{8} + \frac{b}{4}\right) \cdot CO_2$$

$$(1.5)$$

This equation is derived by balancing the total conversion of the organic material to CH_4 and CO_2 with H_2O as the only external source under anaerobic conditions [40]. Such equation can be improved including nitrogen and sulfur, that are entirely converted to NH_3 and H_2S and are assumed to negatively affect anaerobic digestion trough bacteria inhibition and pH variation [41]. Therefore, a general equation is obtained for generic organic matter with formula $C_n H_a O_b N_c S_d$:

$$C_n H_a O_b N_c S_d + \left(n - \frac{a}{4} - \frac{b}{2} + \frac{3c}{4} + \frac{d}{2}\right)H_2O \rightarrow \left(\frac{n}{2} + \frac{a}{8} - \frac{b}{4} - \frac{3c}{8} - \frac{d}{4}\right) \cdot CH_4$$

$$+ \left(\frac{n}{2} - \frac{a}{8} + \frac{b}{4} + \frac{3c}{8} + \frac{d}{8}\right) \cdot CO_2 + c \cdot NH_3 + d \cdot H_2S \qquad (1.6)$$

The theoretical methane yield TBMP, can be obtained from this reaction with the empirical formula:

$$TBMP_{el} = \frac{\left(\frac{n}{2} + \frac{a}{8} - \frac{b}{4} - \frac{3c}{8} - \frac{d}{4}\right) \cdot 22415}{12n + a + 16b + 14c + 32d} \tag{1.7}$$

where the term 22,415 is a volume conversion coefficient that allows to obtain the desired unit of measure of $[mL_{CH_4}/g_{VS}]$.

This formula doesn't account for the content of non-degradable components such as lignin, considering the substrate as completely degradable. The actual methane yield then is lower than the one evaluated from the TBMP.

1.10 Organic Fractions Composition

Another method for the calculation of the TBMP takes advantage of the knowledge of the organic fractions of lipids, carbohydrates, and proteins [42] for the calculation of the TBMP:

$$TBMP_{org} = 415 \cdot \%Carbohydrates + 496 \cdot \%Proteins + 1014 \cdot \%Lipids \tag{1.8}$$

The coefficients in this equation represent the $TBMP_{el}$ of the average chemical formulas for carbohydrates ($C_6H_{10}O_5$), proteins ($C_5H_7O_2N$) and lipids ($C_{57}H_{104}O_6$), calculated with the Buswell formula. The percentage of the macromolecules are in %VS (Volatile Solids) unit. Carbohydrates term is evaluated, if data are available, as the sum of cellulose, hemicellulose, sugar and lignin.

The drawbacks of this formula are that the fractions of carbohydrates, lipids and proteins must be quantified by analytical composition analysis. The calculation does not consider the lost in performance caused by the presence of low-degradable organic matter.

Generally, the elemental analysis method gives slightly higher results with respect to the ones calculated with the organic fractions analysis method, however it could be said that both methods give acceptable results for an estimate [36].

1.11 Experimental Biomethane Potential

The BMP of substrates can be experimentally calculated by performing batch tests. The BMP calculated with such assays represents the ultimate cumulative methane yield that one or more substrates give in optimal conditions, using appropriate solids concentrations and substrate/inoculum ratios to create the best environment for methane production, as well as appropriate digestion times. Despite the usefulness and inexpensiveness of BMP tests, they don't follow a universally accepted protocol and they aren't currently standardized, so they might give quite different results even for the same type of substrate. The variabilities that can be observed in BMP tests are due to possible differences in operating conditions, the substrate/inoculum ratio, the inoculum source, the substrate concentration, and composition; moreover, the experimental setups might differ on their gas measurement and analytical techniques. Therefore, the resulting EBMPs from these tests should be carefully compared and discussed [43]. Despite this, BMP tests allow to obtain approximate, but reliable, values of the methane yield in real conditions.

Example 1.1 Evaluate the TBMP knowing from the ultimate analysis that has been done on cattle manure.

Name	Massive fraction (wt%)
Carbon	11.30
Hydrogen	1.26
Oxygen	8.79
Nitrogen	1.27
Sulphur	0.27

Resolution

Knowing that the ultimate analysis also accounts for halides and other materials, it is necessary to normalize these fractions and convert them into molar ones.

Name	Normalized fraction (wt%)	Molecular weight (kg/kmol)	Molar fraction (mol%)
Carbon	49,3	12	33,0
Hydrogen	5,50	1	44,2
Oxygen	38,4	16	19,3
Nitrogen	5,55	14	3,2
Sulphur	1,18	32	0,3

Now, through Eqs. 1.6 and 1.7 it is possible to evaluate a general biomass degradation to methane and thus, calculate the theoretical biomethane potential. In the CHONS brute formula (1.6), molar values should be approximated or kept.

$C_n H_a O_b N_c S_d \rightarrow C_{33} H_{44.2} O_{19.3} N_{3.2} S_{0.3}$	$\left(n - \frac{a}{4} - \frac{b}{2} + \frac{3c}{4} + \frac{d}{2}\right) = 14.85$
$n = 33$	$\left(\frac{n}{2} + \frac{a}{8} - \frac{b}{4} - \frac{3c}{8} - \frac{d}{4}\right) = 15.93$
$a = 44.2$	$\left(\frac{n}{2} - \frac{a}{8} + \frac{b}{4} + \frac{3c}{8} + \frac{d}{8}\right) = 17.04$
$b = 19.3$	
$c = 3.2$	
$d = 0.3$	

Thus, the final chemical equation become:

$$C_{33}H_{44.2}O_{19.3}N_{3.2}S_{0.3} + 14.85H_2O \rightarrow 15.93CH_4 + 17.04CO_2 + 3.2NH_3 + 0.3H_2S$$

And the TBMP results as:

$$TBMP_{el} = \frac{\left(\frac{n}{2} + \frac{a}{8} - \frac{b}{4} - \frac{3c}{8} - \frac{d}{4}\right) \cdot 22415}{12n + a + 16b + 14c + 32d} = \frac{356958.9}{803.4} = 444.31\frac{mL_{CH_4}}{g_{VS}}$$

Do It Yourself

Knowing from the proximate analysis the relative amount of macromolecules composing the organic matter of cattle manure, as percentage of volatile solids (%VS), calculate the TBMP and the biodegradability of the biomass (BD_1), and compare it with the previous exercise (BD_2).

Cellulose	%VS	16.1	Sugar	%VS	30.286	Lipids	%VS	9.8
Hemicellulose	%VS	13.62	Proteins	%VS	14	Lignin	%VS	9.357
EBMP	%VS	261						

Results: TBMP = 455.57 mL_{CH_4}/g_{VS}; $BD_1 = 0.573$; $BD_2 = 0.587$.

References

1. Rajput AA, Sheikh Z (2019) Effect of inoculum type and organic loading on biogas production of sunflower meal and wheat straw. Sustain Environ Res 29(1):4. https://doi.org/10.1186/s42834-019-0003-x
2. Yarberry A, Lansing S, Luckarift H, Diltz R, Mulbry W, Yarwood S (2019) Effect of anaerobic digester inoculum preservation via lyophilization on methane recovery. Waste Manag 87:62–70. https://doi.org/10.1016/j.wasman.2019.01.033
3. Raj T, Chandrasekhar K, Naresh-Kumar A, Kim S-H (2022) Lignocellulosic biomass as renewable feedstock for biodegradable and recyclable plastics production: a sustainable approach. Renew Sustain Energy Rev 158:112130. https://doi.org/10.1016/j.rser.2022.112130
4. Ebrahimi M et al (2022) Effects of lignocellulosic biomass type on nutrient recovery and heavy metal removal from digested sludge by hydrothermal treatment. J Environ Manage 318:115524. https://doi.org/10.1016/j.jenvman.2022.115524
5. Steffen R, Szolar O, Braun R (1998) Feedstocks for anaerobic digestion. Inst Agrobiotechnol Tulin Univ Agric Sci Vienna 14:1–29
6. Legner M, McMillen DR, Cvitkovitch DG (2019) Role of dilution rate and nutrient availability in the formation of microbial biofilms. Front Microbiol 10:916. https://doi.org/10.3389/fmicb.2019.00916
7. Agawin NSR, Duarte CM, Agustí S (2000) Nutrient and temperature control of the contribution of picoplankton to phytoplankton biomass and production. Limnol Oceanogr 45(3):591–600. https://doi.org/10.4319/lo.2000.45.3.0591

8. Tokuşoglu Ö, Unal MK (2003) Biomass nutrient profiles of three microalgae: spirulina platensis, chlorella vulgaris, and Isochrisis galbana. J Food Sci 68(4):1144–1148. https://doi.org/10.1111/j.1365-2621.2003.tb09615.x
9. Michalik M, Wilczyńska-Michalik W (2012) Mineral and chemical composition of biomass ash. https://doi.org/10.13140/2.1.4298.5603
10. Gil A, Toledo M, Siles JA, Martín MA (2018) Multivariate analysis and biodegradability test to evaluate different organic wastes for biological treatments: anaerobic co-digestion and co-composting. Waste Manag 78:819–828. https://doi.org/10.1016/j.wasman.2018.06.052
11. Xu F, Wang Z-W, Tang L, Li Y (2014) A mass diffusion-based interpretation of the effect of total solids content on solid-state anaerobic digestion of cellulosic biomass. Bioresour Technol 167:178–185. https://doi.org/10.1016/j.biortech.2014.05.114
12. Ghimire A et al (2018) Effect of total solids content on biohydrogen production and lactic acid accumulation during dark fermentation of organic waste biomass. Bioresour Technol 248:180–186. https://doi.org/10.1016/j.biortech.2017.07.062
13. Oosterkamp WJ (2014) Chapter 13: use of volatile solids from biomass for energy production. In: Gupta VK, Tuohy MG, Kubicek CP, Saddler J, Xu F (eds) Bioenergy research: advances and applications. Elsevier, Amsterdam, pp 203–217
14. Ahmadi-Pirlou M, Ebrahimi-Nik M, Khojastehpour M, Ebrahimi SH (2017) Mesophilic co-digestion of municipal solid waste and sewage sludge: effect of mixing ratio, total solids, and alkaline pretreatment. Int Biodeterior Biodegrad 125:97–104. https://doi.org/10.1016/j.ibiod.2017.09.004
15. Jain S, Jain S, Wolf IT, Lee J, Tong YW (2015) A comprehensive review on operating parameters and different pretreatment methodologies for anaerobic digestion of municipal solid waste. Renew Sustain Energy Rev 52:142–154. https://doi.org/10.1016/j.rser.2015.07.091
16. Motte J-C et al (2013) Total solids content: a key parameter of metabolic pathways in dry anaerobic digestion. Biotechnol Biofuels 6(1):164. https://doi.org/10.1186/1754-6834-6-164
17. Meegoda JN, Li B, Patel K, Wang LB (2018) A review of the processes, parameters, and optimization of anaerobic digestion. Int J Environ Res Public Health 15(10):2224. https://doi.org/10.3390/ijerph15102224
18. Siddique MNI, Wahid ZA (2018) Achievements and perspectives of anaerobic co-digestion: a review. J Clean Prod 194(1):359–371. https://doi.org/10.1016/j.jclepro.2018.05.155
19. Wang X, Yang G, Feng Y, Ren G, Han X (2012) Optimizing feeding composition and carbon-nitrogen ratios for improved methane yield during anaerobic co-digestion of dairy, chicken manure and wheat straw. Bioresour Technol 120:78–83. https://doi.org/10.1016/j.biortech.2012.06.058
20. Dima AD, Pârvulescu OC, Mateescu C, Dobre T (2020) Optimization of substrate composition in anaerobic co-digestion of agricultural waste using central composite design. Biomass Bioenergy 138:8–10. https://doi.org/10.1016/j.biombioe.2020.105602
21. Wang M, Li W, Li P, Yan S, Zhang Y (2017) An alternative parameter to characterize biogas materials: available carbon-nitrogen ratio. Waste Manag 62:76–83. https://doi.org/10.1016/j.wasman.2017.02.025
22. Turner BL (2010) Variation in pH optima of hydrolytic enzyme activities in tropical rain forest soils. Appl Environ Microbiol 76(19):6485–6493. https://doi.org/10.1128/AEM.00560-10
23. Sarkar O, Rova U, Christakopoulos P, Matsakas L (2021) Influence of initial uncontrolled pH on acidogenic fermentation of brewery spent grains to biohydrogen and volatile fatty acids production: Optimization and scale-up. Bioresour Technol 319:124233. https://doi.org/10.1016/j.biortech.2020.124233
24. Zhang B, Zhang LL, Zhang SC, Shi HZ, Cai WM (2005) The influence of pH on hydrolysis and acidogenesis of kitchen wastes in two-phase anaerobic digestion. Environ Technol 26(3):329–339. https://doi.org/10.1080/09593332608618563
25. Tang J, Wang XC, Hu Y, Zhang Y, Li Y (2017) Effect of pH on lactic acid production from acidogenic fermentation of food waste with different types of inocula. Bioresour Technol 224:544–552. https://doi.org/10.1016/j.biortech.2016.11.111

26. Cheah Y-K, Vidal-Antich C, Dosta J, Mata-Álvarez J (2019) Volatile fatty acid production from mesophilic acidogenic fermentation of organic fraction of municipal solid waste and food waste under acidic and alkaline pH. Environ Sci Pollut Res 26(35):35509–35522. https://doi.org/10.1007/s11356-019-05394-6
27. Marquart KA et al (2019) Influence of pH on the balance between methanogenesis and iron reduction. Geobiology 17(2):185–198. https://doi.org/10.1111/gbi.12320
28. Kainthola J, Kalamdhad AS, Goud VV (2020) Optimization of process parameters for accelerated methane yield from anaerobic co-digestion of rice straw and food waste. Renew Energy 149:1352–1359. https://doi.org/10.1016/j.renene.2019.10.124
29. Labatut RA, Pronto JL (2018) Chapter 4: sustainable waste-to-energy technologies: anaerobic digestion. In: Trabold TA, Babbitt CW (eds) Sustainable food waste-to-energy systems. Academic Press, New York, pp 47–67
30. Li D et al (2015) Effects of feedstock ratio and organic loading rate on the anaerobic mesophilic co-digestion of rice straw and cow manure. Bioresour Technol 189:319–326. https://doi.org/10.1016/j.biortech.2015.04.033
31. Jabeen M, Zeshan S, Yousaf S, Haider MR, Malik RN (2015) High-solids anaerobic co-digestion of food waste and rice husk at different organic loading rates. Int Biodeterior Biodegrad 102:149–153. https://doi.org/10.1016/j.ibiod.2015.03.023
32. Labatut RA, Angenent LT, Scott NR (2011) Biochemical methane potential and biodegradability of complex organic substrates. Bioresour Technol 102(3):2255–2264. https://doi.org/10.1016/j.biortech.2010.10.035
33. Nielfa A, Cano R, Fdz-Polanco M (2015) Theoretical methane production generated by the co-digestion of organic fraction municipal solid waste and biological sludge. Biotechnol Rep 5(1):14–21. https://doi.org/10.1016/j.btre.2014.10.005
34. Aboudi K, Álvarez-Gallego CJ, Romero-García LI (2017) Influence of total solids concentration on the anaerobic co-digestion of sugar beet by-products and livestock manures. Sci Total Environ 586:438–445. https://doi.org/10.1016/j.scitotenv.2017.01.178
35. Varsha SSV, Soomro AF, Baig ZT, Vuppaladadiyam AK, Murugavelh S, Antunes E (2020) Methane production from anaerobic mono- and co-digestion of kitchen waste and sewage sludge: synergy study on cumulative methane production and biodegradability. Biomass Convers Biorefinery 12:3911–3919. https://doi.org/10.1007/s13399-020-00884-x
36. Li Y, Zhang R, Liu G, Chen C, He Y, Liu X (2013) Comparison of methane production potential, biodegradability, and kinetics of different organic substrates. Bioresour Technol 149:565–569. https://doi.org/10.1016/j.biortech.2013.09.063
37. Triolo JM, Sommer SG, Møller HB, Weisbjerg MR, Jiang XY (2011) A new algorithm to characterize biodegradability of biomass during anaerobic digestion: influence of lignin concentration on methane production potential. Bioresour Technol 102(20):9395–9402. https://doi.org/10.1016/j.biortech.2011.07.026
38. Sawyerr N, Trois C, Workneh T (2019) Identification and characterization of potential feedstock for biogas production in South Africa. J Ecol Eng 20(6):103–116. https://doi.org/10.12911/22998993/108652
39. Buswell AM, Mueller HF (1952) Mechanism of methane fermentation. Ind Eng Chem 44(3):550–552. https://doi.org/10.1021/ie50507a033
40. Angelidaki I, Sanders W (2004) Assessment of the anaerobic biodegradability of macropollutants. Rev Environ Sci Biotechnol 3(2):117–129. https://doi.org/10.1007/s11157-004-2502-3
41. Hidalgo D, Martín-Marroquín JM (2015) Biochemical methane potential of livestock and agri-food waste streams in the Castilla y León Region (Spain). Food Res Int 73:226–233. https://doi.org/10.1016/j.foodres.2014.12.044
42. Raposo F et al (2011) Biochemical methane potential (BMP) of solid organic substrates: evaluation of anaerobic biodegradability using data from an international interlaboratory study. J Chem Technol Biotechnol 86(8):1088–1098. https://doi.org/10.1002/jctb.2622
43. Ohemeng-Ntiamoah J, Datta T (2019) Perspectives on variabilities in biomethane potential test parameters and outcomes: a review of studies published between 2007 and 2018. Sci Total Environ 664:1052–1062. https://doi.org/10.1016/j.scitotenv.2019.02.088

Chapter 2
Database Introduction

Statistical methods should always be applied in the presence of enough amount of data. In this book, the data used have been taken gathering a conspicuous amount of information from the literature, and resumed in the work of Moretta et al. [1]. Then, these have been collected in a single database, which structure and content are described in the next sections.

2.1 Database Details

The database has 229 entries with 21 features. Four different categories have been chosen to gather substrates with similar origin and characteristics. In particular: Manure,.Agricultural waste, Organic waste, and Sludges. These are subdivided into the relative substrates. In the database it is also possible to find the same substrates many times. This because there can exist many elemental and macro-configurations (i.e., macromolecule content) of the same substrate. Indeed, there exist many analytical methods to measure the biomass composition, and all of them differ in the methodology and analysis [2–4]. Moreover, many data of the same substrates are very useful when conducting machine learning algorithm as classification or clustering. In fact, it is possible to find the actual range of a certain variable (i.e., total organic carbon, protein content, etc.) of that specific biomass. As a consequence, it is possible to exploit the cluster centroid instead of the arithmetical mean for further statistical studies (i.e., normal distribution) [5]. The list of features and their details are reported in Table 2.1.

All the biomasses present in the database, which will be used in the modeling in the next chapters, are listed in the Table 2.2. In the next section, a brief description of the principal properties of these biomasses is presented.

F. Moretta and G. Bozzano, *Mathematical and Statistical Approaches for Anaerobic Digestion Feedstock Optimization*, SpringerBriefs in Energy, https://doi.org/10.1007/978-3-031-56460-4_2

Table 2.1 Database features
and relative details

Name	Description	Unit of measure
TS	Total solids content	%w/w
VS	Volatile solids content	$\%w/w_{ts}$
TOC	Total organic carbon	$\%w/w_{ts}$
C	Carbon content	%mol
H	Hydrogen content	%mol
O	Oxygen content	%mol
N	Nitrogen content	%mol
S	Sulphur content	%mol
CN	Carbon–nitrogen ration	–
LP	Lipids content	$\%w/w_{ts}$
PR	Protein content	$\%w/w_{ts}$
SU	Sugar content	$\%w/w_{ts}$
LG	Lignin content	$\%w/w_{ts}$
CE	Cellulose content	$\%w/w_{ts}$
HCE	Hemicellulose content	$\%w/w_{ts}$
SA	Salts/Ashes	$\%w/w_{ts}$
TBMP	Theoretical biomethane potential	Ml_{ch4}/g_{vs}
BMP	Experimental biomethane potential	Ml_{ch4}/g_{vs}
BD	Biodegradability	–

2.2   Database Biomasses Description

Despite the natural origin of every biomass, the specific content of some of its attributes differ from each other. For all names and symbology please refer to Table 2.2.

2.3   Manure

Dairy Manure is typically collected by a scraper system, and often straw—used as bedding material—is present too, resulting in slight variations of the total solids and lignocellulosic compounds content (lignin, cellulose, and hemicellulose). The high variability in composition, CN, TS and VS content depends on many factors such as the animal housing system, the location, animals' age and diet. The C/N ratio of DM is generally quite low, typical of manures, due to the high proteins content, which also represent one of the main nitrogen carriers. The high lignin, cellulose and hemicellulose content are due to the presence of straw coming from the gathering process, while lipids content is quite low and shows a high variability. The BMP

Table 2.2 Database biomasses names and relative category subdivision

Name	Category	Symbol
Dairy manure	Manure	DM
Pig manure	Manure	PM
Sow manure	Manure	SM
Chicken manure	Manure	CM
Chicken litter	Manure	CL
Sheep manure	Manure	SHM
Goat manure	Manure	GM
Straw	Agricultural waste	ST
Wheat straw	Agricultural waste	WS
Rice straw	Agricultural waste	RS
Rice husk	Agricultural waste	RH
Sugar-beet byproduct	Agricultural waste	SBB
Dry grass	Agricultural waste	DG
Corn stover	Agricultural waste	CS
Potato waste	Agricultural waste	PW
Yard waste	Agricultural waste	YW
Vinegar residue	Agricultural waste	VR
Food waste	Organic waste	FW
Fruit and vegetable food waste	Organic waste	FVFW
Organic fraction municipal solid waste	Organic waste	OFMSW
Fish waste	Organic waste	FSW
Slaughterhouse residue	Organic waste	SR
Blood	Organic waste	BL
Pre-cocked waste	Organic waste	PCW
Exhaust kitchen oil	Organic waste	EKO
Sewage sludge	Sludges	SS
Food industry sludge	Sludges	FIS

is extremely variable since it depends on the composition and on the batch assay's experimental method [6].

On the other hand, Pigs [7] are usually kept in feedlots with open floors where the excrements are collected through slots or scraper systems [8]. According to the housing system, the TS is highly variable, and PM can be collected both as a liquid slurry (TS $= 1 \div 5\%$) or with a higher dry matter content. Similarly, also all the other parameters may change according to the same factors mentioned in the previous paragraph. As all manures, the C/N ratio of PM is very low, indeed the proteins content is quite high. Also in this case, lignin, cellulose, and hemicellulose contents are quite high due to the presence of lignocellulosic compounds such as straw coming from the housing system. The lipids content is also quite high and the BMP is generally higher than for DM.

Sows' Manure [9], especially when gestating, is usually distinguished from Pig Manure due to some differences in their composition and on their methane production, which is usually lower.

C/N ratio values are reasonable and similar to other manures.

Chickens are usually kept in open feedlots holding up to several hundred thousand animals. Keeping these in open feedlots typically causes a significant contamination of manure with sand and bedding materials such as straw [8], then CM is usually characterised by high TS contents (20 ÷ 50%) and high quantities of ash and ligno-cellulosic compounds. The C/N ratios are the lowest among all the manures: this is because CM is characterised by high concentrations of proteins. The lipids content is generally the lowest among manures, and the EBMP is quite low as well [10].

Compared to chicken manure, chicken litter contains a higher quantity of bedding materials such as straw and wood shavings, which are typical lignocellulosic matters. Therefore, CL contains more carbon and less nitrogen with respect to CM and can be classified as an independent substrate. The composition of CL may considerably vary according to the bedding materials and animals' diet. TS is generally very high, as a consequence of the high values of C/N ratio, lignin content, cellulose content and hemicellulose content. On the other hand, the lipids and proteins contents are quite similar than those of CM. The higher quantity of lignocellulosic compounds leads to a lower mean value of the BMP, and thus a lower biodegradability [10].

Sheep Manure, as all manures, is characterised by a low C/N ratio and by the presence of bedding materials, which lead to a quite high TS and lignin content.

Goat Manure is generally characterised by quite high TS (from about 35% to 80%), lower C/N ratios and lignin content, which stresses the fact that GM usually contains high quantities of bedding materials [11].

2.4 Agricultural Waste

One of the principal agricultural waste is the straw [12–14]. It comprises different species, particularly Wheat Straw and Rice Straw, which have similar characteristics and can be generalised in the same category.

Since straw is a dry substrate, it is generally characterised by a very high TS and high organic content. The percentages of lipids and proteins are usually low, and this leads to higher C/N ratios. Generally, the principal components are lignin, cellulose, and hemicellulose. Since the lignocellulosic content is very high, the BMP are usually low, and highly depend on the operating conditions of the batch test and on eventual pre-treatments.

Rice Husk [15] is a very interesting biomass source, since it is an abundant by-product of rice cultivation and is characterised by a very low degradability, then it could be interesting to see how its methane production would improve with co-digestion. Similarly, to straw, the TS content is very high, and the major components are lignin, cellulose and hemicellulose. Furthermore, being the proteins content very low, the C/N ratio can reach very high values.

Sugar-beet [16] waste comprises all the by-products that come from sugar beets cultivation and processing, like roots, leaves, exhausted pulp, and molasses. According to the composition of this substrate, its parameters can considerably change: for example, the TS content and the C/N ratio are highly variable. The BMP of this category are usually very high.

About dry grass [17], different types have been considered in the database, such as switchgrass, meadow grass and hay grass, which are characterised by similar compositions. Similarly, to straw, dry grass is characterised by high TS, high C/N ratios, high quantities of lignocellulosic compounds and quite low BMP. The C/N ratio is usually about 60, and the main components are generally cellulose and hemicellulose, while lignin content is quite low in this case.

Corn Stover [18] derives from maize cultivation too, and consists of dried leaves, stalks and cobs of maize plants. This makes it a lignocellulosic substrate similar to straw. Indeed, the C/N ratio is usually high, and the lignin, cellulose and hemicellulose contents are high as well.

Potato waste comprises mainly fresh potato residues and potato peels. The C/N ratio is usually comprised between 30 and 40, then the BMP are generally quite high.

Yard Waste [19] comprises house-garden grass and plants, leaves, tree trimming and harvest remains. Usually the C/N ratio is high, and a high quantity of hardly degradable matter is contained, then the BMP is usually quite low.

Vinegar Residue [20] is a typical solid by-product of the vinegar production process. It is characterised by a very high C/N ratio and a high content of lignocellulosic compounds.

2.5   Organic Waste

Food Waste includes all the animal and vegetable food wastes coming from houses, kitchens and supermarkets; thus, it is a very abundant substrate. FW contains a high quantity of readily biodegradable organic materials; consequently, anaerobic digestion is an ideal solution for its energy recovery. Indeed, it usually contains a high amount of lipids and proteins, which are easily degradable compounds, and a quite low quantity of lignin, cellulose and hemicellulose. The C/N ratio, however, is usually quite low and comprised between about 16 and 22 depending on the composition. Thanks to the high organic content the BMP is usually high, making this substrate one of the most promising for co-digestion with hardly degradable material [15].

Fruit and vegetable food waste includes food waste of vegetable origin only. With respect to FW, the lipids content of FVFW is lower due to the absence of animal-based food residues, and the C/N ratio is often lower than FW. The BMP, then, is generally lower than the one of FW [21].

The OFMSW represents a household waste that generally includes food waste, garden waste, paper and textile residues. Its composition varies depending on a range of factors, reflecting the population density, the location and season of production. Globally, an enormous amount of OFMSW is generated, most of which is burnt or landfilled; its uncontrolled decomposition contributes to climate change and pollution of soil, water and air. Thanks to its high organic content anaerobic digestion is an environmentally friendly solution to recover OFMSW, preventing its related pollution.

Due to the extreme variability in its composition, all the parameters characterising OFMSW can considerably vary depending on its source: C/N ratio may vary from low values such as 10 to relatively high values as 50, then an average value with a certain distribution will be defined; the TS may considerably change depending on the moisture content of the waste and on the organic composition. The BMP is usually quite high due to the presence of a high amount of organic matter, however it is meanly lower than the one of FW since in this case the substrate contains a higher amount of hardly degradable fractions (e.g., lignin content) [22, 23].

Fish Waste consists of residues coming from fish processing, such as heads, tails, fish bones and viscera. This substrate contains a high quantity of lipids, and despite the high proteins content, it shows quite high C/N ratios, that fall into the optimal range of $20 \div 40$, and the lignocellulosic content is generally low. The BMP is generally high, which makes this substrate very promising for this kind of process [24].

Slaughterhouse Residue represents the residues coming from meat processing— for example, viscera, intestine and digestive tract contents—of animals such as pigs, cows, chicken, sheep, goats, and rabbits. The C/N ratio may vary depending on the meat's nature, and it is generally quite low. This can be explained considering that lipids and proteins content is generally high. A high contents of lignin, cellulose and hemicellulose are observed too, probably due to the presence of lignocellulosic bedding materials [25].

Blood is a slaughterhouse residue as well, but it is considered separately due to some characteristics that distinguish it from other slaughterhouse wastes. The C/N ratio of blood is extremely low; this because the most abundant component of Blood are proteins, which can even reach the $96.5\% w/w_{TS}$. Despite the C/N value, the BMP is relatively high [26].

Pre-cocked waste involves precooked starchy food, such as pasta, generally rejected from the quality control line. One of the principal macronutrients is protein, followed by lipids, while no lignocellulosic compounds are present. As a consequence, BMP is generally quite high [23, 27, 28].

Exhaust kitchen oil represents exhaust oil coming from kitchens of both animal and vegetable origin. EKO is mainly composed of lipids, that represent nearly the $100\% w/w_{TS}$, and so the C/N ratio is usually quite high [29–31]. Due to these characteristics, the BMP is extremely high, then this substrate represents a sort of *"outlier"* with respect to the other substrates, and it won't be considered on the statistical analysis.

2.6 Sludges

Sewage Sludge represents the active sludge coming from wastewater treatment plants after decanting processes. This substrate is extremely abundant of pathogens such as viruses and bacteria, therefore it needs to be treated before being disposed and used, for example, as fertilizer. The quite high organic and microorganism content allows it to be suitable for anaerobic digestion, even if C/N ratio is usually low and it is highly contaminated by heavy metals and salts. Due to this, generally the BMP is not so high. Its composition is very variable, it depends on the sludge source [8, 16, 18, 19].

Finally, food industry sludge can be distinguished from SS since it derives from the treatment of food industry wastewaters [7]. Since, the quantity and the quality of the information collected about this substrate is quite low, then it won't be considered for further analysis.

References

1. Moretta F, Goracci A, Manenti F, Bozzano G (2022) Data-driven model for feedstock blending optimization of anaerobic co-digestion by BMP maximization. J Clean Prod 375:134140. https://doi.org/10.1016/j.jclepro.2022.134140
2. Jr SV (2014) Analytical techniques for the chemical analysis of plant biomass and biomass products. Anal Methods 6(20):8094–8105. https://doi.org/10.1039/C4AY00388H
3. Karimi K, Taherzadeh MJ (2016) A critical review of analytical methods in pretreatment of lignocelluloses: composition, imaging, and crystallinity. Bioresour Technol 200:1008–1018. https://doi.org/10.1016/j.biortech.2015.11.022
4. Achinas S, Euverink GJW (2016) Theoretical analysis of biogas potential prediction from agricultural waste. Resour Effic Technol 2(3):143–147. https://doi.org/10.1016/j.reffit.2016.08.001
5. Cluster Centroid: an overview|ScienceDirect Topics. Consultato: 19 luglio 2023. Disponibile su: https://www.sciencedirect.com/topics/computer-science/cluster-centroid
6. Page DI, Hickey KL, Narula R, Main AL, Grimberg SJ (2008) Modeling anaerobic digestion of dairy manure using the IWA anaerobic digestion model no. 1 (ADM1). Water Sci Technol 58(3):689–695. https://doi.org/10.2166/wst.2008.678
7. Murto M, Björnsson L, Mattiasson B (2004) Impact of food industrial waste on anaerobic co-digestion of sewage sludge and pig manure. J Environ Manage 70(2):101–107. https://doi.org/10.1016/j.jenvman.2003.11.001
8. Steffen R, Szolar O, Braun R (1998) Feedstocks for anaerobic digestion. Inst Agrobiotechnol Tulin Univ Agric Sci Vienna, pp 1–29
9. Bywater A, Kusch-Brandt S (2022) Exploring farm anaerobic digester economic viability in a time of policy change in the UK. Processes 10(2):2. https://doi.org/10.3390/pr10020212
10. US EPA O (2023) Anaerobic digestion on poultry farms. Consultato: 19 luglio 2023. Disponibile su: https://www.epa.gov/agstar/anaerobic-digestion-poultry-farms
11. Li Y, Achinas S, Zhao J, Geurkink B, Krooneman J, Euverink GJW (2020) Co-digestion of cow and sheep manure: performance evaluation and relative microbial activity. Renew Energy 153:553–563. https://doi.org/10.1016/j.renene.2020.02.041
12. Kainthola J, Kalamdhad AS, Goud VV (2020) Optimization of process parameters for accelerated methane yield from anaerobic co-digestion of rice straw and food waste. Renew Energy 149:1352–1359. https://doi.org/10.1016/j.renene.2019.10.124

13. Li D et al (2015) Effects of feedstock ratio and organic loading rate on the anaerobic mesophilic co-digestion of rice straw and cow manure. Bioresour Technol 189:319–326. https://doi.org/10.1016/j.biortech.2015.04.033

14. Wang X, Yang G, Feng Y, Ren G, Han X (2012) Optimizing feeding composition and carbon–nitrogen ratios for improved methane yield during anaerobic co-digestion of dairy, chicken manure and wheat straw. Bioresour Technol 120:78–83. https://doi.org/10.1016/j.biortech.2012.06.058

15. Jabeen M, Zeshan S, Yousaf S, Haider MR, Malik RN (2015) High-solids anaerobic co-digestion of food waste and rice husk at different organic loading rates. Int Biodeterior Biodegrad 102:149–153. https://doi.org/10.1016/j.ibiod.2015.03.023

16. Anaerobic digestion of sugar beet pulp after acid thermal and alkali thermal pretreatments|SpringerLink. Consultato: 19 luglio 2023. Disponibile su: https://doi.org/10.1007/s13399-019-00539-6

17. Koch K, Lübken M, Gehring T, Wichern M, Horn H (2010) Biogas from grass silage: measurements and modeling with ADM1. Bioresour Technol 101(21):8158–8165. https://doi.org/10.1016/j.biortech.2010.06.009

18. Croce S, Wei Q, D'Imporzano G, Dong R, Adani F (2016) Anaerobic digestion of straw and corn stover: the effect of biological process optimization and pre-treatment on total bio-methane yield and energy performance. Biotechnol Adv 34(8):1289–1304. https://doi.org/10.1016/j.biotechadv.2016.09.004

19. Gupta P, Singh RS, Sachan A, Vidyarthi AS, Gupta A (2012) Study on biogas production by anaerobic digestion of garden-waste. Fuel 95:495–498. https://doi.org/10.1016/j.fuel.2011.11.006

20. Li L et al (2015) Anaerobic digestion performance of vinegar residue in continuously stirred tank reactor. Bioresour Technol 186:338–342. https://doi.org/10.1016/j.biortech.2015.03.086

21. Xu F, Li Y, Ge X, Yang L, Li Y (2018) Anaerobic digestion of food waste: challenges and opportunities. Bioresour Technol 247:1047–1058. https://doi.org/10.1016/j.biortech.2017.09.020

22. Ahmadi-Pirlou M, Ebrahimi-Nik M, Khojastehpour M, Ebrahimi SH (2017) Mesophilic co-digestion of municipal solid waste and sewage sludge: effect of mixing ratio, total solids, and alkaline pretreatment. Int Biodeterior Biodegrad 125:97–104. https://doi.org/10.1016/j.ibiod.2017.09.004

23. Jain S, Jain S, Wolf IT, Lee J, Tong YW (2015) A comprehensive review on operating parameters and different pretreatment methodologies for anaerobic digestion of municipal solid waste. Renew Sustain Energy Rev 52:142–154. https://doi.org/10.1016/j.rser.2015.07.091

24. Bücker F et al (2020) Fish waste: an efficient alternative to biogas and methane production in an anaerobic mono-digestion system. Renew Energy 147:798–805. https://doi.org/10.1016/j.renene.2019.08.140

25. Selormey GK, Barnes B, Kemausuor F, Darkwah L (2021) A review of anaerobic digestion of slaughterhouse waste: effect of selected operational and environmental parameters on anaerobic biodegradability. Rev Environ Sci Biotechnol 20(4):1073–1086. https://doi.org/10.1007/s11157-021-09596-8

26. Energies|Free Full-Text|Anaerobic digestion of blood from slaughtered livestock: a review. Consultato: 19 luglio 2023. Disponibile su: https://www.mdpi.com/1996-1073/14/18/5666

27. Naveen CC et al (2022) Effects of different parameters and co-digestion options on anaerobic digestion of parboiled rice mill wastewater: a review. BioEnergy Res. https://doi.org/10.1007/s12155-022-10522-1

28. Kazemi-Bonchenari M, Alizadeh A, Javadi L, Zohrevand M, Odongo NE, Salem AZM (2017) Use of poultry pre-cooked slaughterhouse waste as ruminant feed to prevent environmental pollution. J Clean Prod 145:151–156. https://doi.org/10.1016/j.jclepro.2017.01.066

29. Carnevale E, Molari G, Vittuari M (2017) Used cooking oils in the biogas chain: a technical and economic assessment. Energies 10:192. https://doi.org/10.3390/en10020192

30. Yan W, Vadivelu V, Maspolim Y, Zhou Y (2021) In-situ alkaline enhanced two-stage anaerobic digestion system for waste cooking oil and sewage sludge co-digestion. Waste Manag 120:221–229. https://doi.org/10.1016/j.wasman.2020.11.047

31. Tamborrino A, Catalano F, Leone A, Bianchi B (2021) A real case study of a full-scale anaerobic digestion plant powered by olive by-products. Foods 10(8):8. https://doi.org/10.3390/foods10081946

Chapter 3
Statistical Analysis

It is important to discover the connections undergoing between data in order to asses complete and robust mathematical relations [1].

There exist many different methodologies. It is, however, necessary to look for the ones which are simple and effective together, in order to quickly progress and reduce the probability of error.

In this chapter the relations between the biomass attributes in the database will be studied, analysing how they distributes and how to exploit these results to find more information and interrelations between each group of data.

Since the number of combination and analysis performed is huge, all the results and related results are available in the drive folder of the database and will be always up-to-date together with the database itself. For the sake of simplicity, only the principal and most relevant ones will be reported.

3.1 Biomass Attributes Distributions

As stated by the name, data distribution asses how values are distributed for a specific field. Data distribution gives information about the median, the mean and the variance of a specific set of values, which can be further used as instruments to evaluate data goodness and correlations through other specific methods. Furthermore, it is possible to use this analysis to estimate the probability of any specific observation to be in a specific *data space*, or in other words, the probability of occurrence of the observation, or estimation, in the data range. This is linked to the confidence interval, which is an estimation percentage value of range which take uncertainty into account to tell how much data is most likely to be around the mean value. The confidence interval at 95% can be mathematically stated with the following expression:

$$\mu - 1.96\frac{\sigma}{\sqrt{n}} \leq \mu \leq \mu + 1.96\frac{\sigma}{\sqrt{n}} \tag{3.1}$$

where μ is the arithmetic mean of the data under study, σ is their standard deviation and n is the population number. These parameters can be estimated as follows, where x_i indicates the single datum from the chosen field:

$$\mu = \frac{1}{n}\sum_{i=1}^{n} x_i \tag{3.2}$$

$$\sigma = \sqrt{\frac{1}{1-n}\sum_{i=1}^{n}(x_i - \mu)^2} \tag{3.3}$$

For a practical example, Fig. 3.1 report the density distribution of two attributes of two different catègory. In particular, Fig. 3.1a shows the histograms of CN value for Manure, and Fig. 3.1b shows the histograms of the HCE content for Agricultural waste. In this figure, and so in the next ones, the lighter colour (yellow) denotes a higher density value, while the darker ones (green and blue, respectively) denote lower density values.

Mean, confidence interval and deviation values for this analysis are reported in Table 3.1. Generally, the data relative to the biomass are really heterogenous, with a relatively high standard deviation. This trend is confirmed by this analysis. The information about how data distribute allow to judge the prediction reliability of a model. In fact, the confidence interval reveals in which range one should expect the values computed. In this particular case: 11.44 and 14.22 for Manure CN, 22.18 and 27.64 for Agricultural waste HCE. So, it is expected that a good model lays its evaluation inside these ranges.

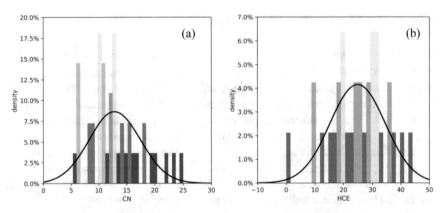

Fig. 3.1 **a** distribution of CN values for Manure category; **b** distribution of HCE content [%w/w$_{TS}$] for Agricultural Waste category

Table 3.1 Statistical results after analysis from data reported in Fig. 3.1

Category	Attribute	Mean	Standard deviation	95% confidence
Manure	CN	12.83	4.67	± 0.72
Agricultural waste	HCE	24.91	9.75	± 1.72

However, this approach is reliable only when the data population is big enough to reach a certain robustness in the analysis results. Specially for not-normal distribution as in this case. Statistically, a population number bigger 30 is a general rule of thumb to achieve a reliable analysis, and in specific cases, when deviation is not high, also 20 is enough [2].

For example, Fig. 3.2 report two very different scenarios: Fig. 3.2a shows the really narrow distribution of VS content in Organic waste; Fig. 3.2b instead show the distribution of TS content in Sludges. While the analysis performed for VS content is robust, results obtained for TS content analysis in sludge are not reliable. Table 3.2 reports the numerical results of the analysis. The lack of data does not allow the construction of a well performing distribution, and so become much more difficult to assess the wellness of that variable prediction.

With these considerations, only data with population high enough will be used to establish correlations in the dataset.

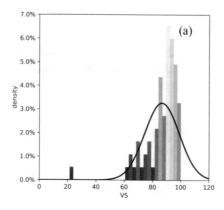 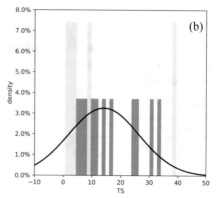

Fig. 3.2 **a** distribution of VS [%w/w$_{TS}$] values for Organic waste category; **b** distribution of TS content [%w/w] for Sludge category

Table 3.2 Statistical results after analysis from data reported in Fig. 3.2

Category	Attribute	Mean	Standard deviation	95% confidence
Organic waste	VS	86.61	12.30	± 1.47
Sludges	TS	14.08	12.55	± 2.74

Example 3.1 Evaluate the mean, standard deviation and confidence interval of protein content in Dairy Manure biomass and compare it to the Chicken Manure. Use data stored in: https://tinyurl.com/28ox9lck.

Resolution

For the mean, it is possible to use (3.2), for standard deviation (3.3) and finally for confidence interval the expression in (3.1). From the database, dairy manure has 15 data, whilst chicken manure has 10 data. The statistical analysis for proteins brings to the following results:

Property	Equation	Dairy manure	Chicken manure
Mean (μ)	$\frac{1}{n}\sum_{i=1}^{n} x_i$	138	205
Standard deviation (σ)	$\sqrt{\frac{1}{1-n}\sum_{i=1}^{n}(x_i - \mu)^2}$	128.2	76.4
Confidence interval	$1.96\frac{\sigma}{\sqrt{n}}$	0.65%	0.47%

Finally, it is possible to say that the protein content in dairy and chicken manure have the following ranges:

Dairy manure	$73 \leq 138 \leq 203$
Chicken manure	$157 \leq 205 \leq 252$

Do It Yourself

Made a complete statistical analysis (Mean, Standard deviation and Confidence interval) of Sugar content in dairy manure and chicken manure, and comment the results obtained.

Results

Dairy manure	$214 \leq 303 \leq 392$
Chicken manure	$65 \leq 134 \leq 203$

3.2   Biomass Attributes Interrelations Evaluation

When working with heterogeneous data as the ones in the database, it is necessary to divide the whole in subgroups that have similar characteristics, in order to increase model precision and thus reducing error propagation [3]. In this case, the database is divided into four subgroups, one for each category: manure, agricultural waste, organic waste, and sludges.

Each category is composed by different kind of biomasses that share origins and characteristics, from which it is possible to find average properties and well estimate the general attributes for each biomass category.

3.3 Linear Model Regression

The simplest and very effective statistical model used for finding data pathways and relations is the linear one [4]. Consequently, the linear regression between each attribute is studied for every biomass category separately. Of course, not all the attributes can have link with each other, therefore only the principal ones, that confirm the presence of a correlation, are shown. To assess the performance of a correlation, the R^2 as performance indicator will be used, and correlation coefficients will be computed to confirm the wellness of every model [5].

Linear regression is performed by minimizing the least square error between the data in the database and values obtained during the training of the model [6]. The aim of this procedure is to find appropriate parameters values which represent at best the data pathway available. Starting from a simple line Eq. (3.4):

$$\hat{y} = \alpha \cdot x + \beta \tag{3.4}$$

where $\hat{y}$ is the predictor, x is the regressor, α and β are the model parameter, namely slope and intercept, respectively. These parameters can be find also using the following relations:

$$\alpha = \frac{n \cdot \sum_i (x_i \cdot y_i) - \sum_i x_i \cdot \sum_i y_i}{n \cdot \sum_i x_i^2 - \left(\sum_i x_i\right)^2} \tag{3.5}$$

$$\beta = \frac{\sum_i y_i \cdot \sum_i x_i^2 - \sum_i x_i \cdot \sum_i (x_i \cdot y_i)}{n \cdot \sum_i x_i^2 - \left(\sum_i x_i\right)^2} \tag{3.6}$$

where n is the number of data used in the regression and y_i is the data available from the database. Finally, the performances are computed using error indicators, and the RMSE is the most used (3.7).

$$RMSE = \sqrt{\frac{\sum_{i=1}^{n}(\hat{y}_i - y_i)^2}{n}} = \sqrt{\frac{\sum_{i=1}^{n}(\alpha \cdot x_i + \beta - y_i)^2}{n}} \tag{3.7}$$

The minimization of this error indicator finds the best α and β that fit the data. The linear models have been implemented using Python™ language, and every pair of attributes studied have been associated with a weight, equal to the reciprocal (3.6) of the squared standard deviation, to account for the data heterogeneity and outlier presence.

$$w = \frac{1}{\sigma^2} \tag{3.8}$$

3.4   Evaluation Metrics

It is important though to assure the wellness of the regression. This can be done exploiting different performance parameter. One of the most used and well-describing the regression goodness is the coefficient of determination R^2_{adj}. R^2 describes how much of the dependent variable's variance that is accounted for by each and every independent variable, whereas R^2_{adj} only accounts for the independent variables that have a direct impact on the dependent ones. R^2 is evaluated by combining other two error indicators: the residual sum of squares (RSS, 3.9) and the total sum of squares (TSS, 3.10):

$$RSS = \sum_{i=1}^{n} (\alpha \cdot x_i + \beta - y_i)^2 \tag{3.9}$$

$$TSS = \sum_{i=1}^{n} (\bar{y} - y_i)^2 \tag{3.10}$$

where $\bar{y}$ is the mean of the predictor values. Evaluation of other error parameters as RMSE and residuals is important since a model can have a good R^2 value but have a significant bias. This is why it is always useful to make use of residual indicators and plots. At this point, the R^2 is computed as one minus the ration between the RSS and TSS (3.11):

$$R^2 = 1 - \frac{RSS}{TSS} \tag{3.11}$$

R^2 tends to rise as the number of independent variables rises. This might be deceptive. As a result, the model is penalized by the R^2_{adj} for including additional independent variables (k in the equation) that do not fit the model (3.12):

$$R^2_{adj} = \left[\frac{(1 - R^2)(n - 1)}{n - k - 1} \right] \tag{3.12}$$

Example 3.2 Find linear regression coefficients between nitrogen content and protein content of the OFMSW and evaluate the model performances. Use data store in: https://tinyurl.com/28ox9lck.

Resolution

By using the nitrogen content as the regressor (x) and the protein content as the predictor (y), it is possible to find the linear correlation coefficients using 3.5 and 3.6. Gathering all the data from the database, the contributions are so computed:

n	$\sum_i x_i$	$\sum_i y_i$	$\sum_i (x_i \cdot y_i)$	$\sum_i x_i^2$	α	β
16	37.43	221.5	586.9	98.96	6.029	-0.259

Thus, the final linear correlation is the following: $PR = 6.029 \cdot N - 0.259$

$N(x)$	$PR(y)$	$x \cdot y$	x^2
1.70	11.99	20.39	2.89
0.67	1.59	1.06	0.44
2.38	11.95	28.41	5.65
3.51	32.02	112.36	12.32
1.42	8.41	11.91	2.01
2.98	10.77	32.12	8.90
1.04	11.38	11.78	1.07
3.11	17.74	55.10	9.65
2.30	13.87	31.96	5.31
2.83	27.82	78.65	7.99
1.84	6.67	12.31	3.40
1.54	8.26	12.68	2.36
2.75	15.00	41.29	7.58
3.12	17.00	53.03	9.73
2.94	16.00	46.98	8.62
3.32	11.12	36.94	11.03

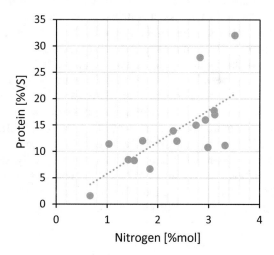

The performances of the regression can be computed through the RMSE evaluation, which using 3.7 gives:

$$RMSE = \sqrt{\frac{\sum_{i=1}^{n}\left(\hat{y}_i - y_i\right)^2}{n}} = \sqrt{\frac{\sum_{i=1}^{n}(6.029 \cdot N_i - 0.259 - PR_i)^2}{16}} = 0.719$$

Showing good correlation performances. On the other hand, the computation of the R^2 is 0.4866. This is not a very high value statistically speaking, but as it is possible to notice from the graph, linear equation successfully represents the correlation between these two sets of data.

Do It Yourself

Perform a linear regression (coefficients, performance indicators) between nitrogen and protein content and for the Fish waste and compare it to the previous exercise.

Results

α	β	RMSE	R^2
6.552	5.346	$6.82 \cdot 10^{-8}$	0.4136

3.5   Statistical Indexes

The linear regression is a good and simple way to find data correlations. However, it is not always the best, or simply it is not enough to the describe what is the actual data relations. Indeed, multilinear of polynomial relations may be used in this manner to retrieve outperforming models. However, a trial-and-error choice of the model type that better fit the data is not sustainable and can be misleading. To help in this procedure, correlations indexes aid to find which are the set of data that better interrelate each other [7, 8]. The main correlation indexes are the Pearson coefficient, the Kendal's rank coefficient and the Spearman's rank coefficient.

The level of linearity between two sets of data is measured by the Pearson correlation coefficient (PCC). It is a normalized measurement of the covariance, with the result always falling between -1 and 1. It denotes the ratio between the covariance of two variables and the product of their standard deviations (3.13). The measure, like covariance itself, can only account for linear correlations between variables and excludes other kinds of interconnections or models. Consequently, PCC can be used to assess if a correlation can be well-described by a linear model or not. The more the value is equal to 0, the more the model in non-linear. When the PCC is equal to 1, the model is perfectly linear, whilst its sign denotes the slope orientation.

$$PCC(X, Y) = \frac{cov(X, Y)}{\sigma_X \sigma_Y} \tag{3.13}$$

where $cov(X, Y)$ is the covariance of the data pair X and Y and σ_i is the standard deviation of the i-th variable.

The Kendall rank correlation coefficient is used to denote if two variables are statistically dependent and to quantify the relationship between two sets, indicating their level of similarity and orderings when ranked. The Kendall correlation between two variables will be high for observations with a similar rank (or identical rank for a correlation of 1) between the two variables, and low for observations with a dissimilar rank (or fully different rank for a correlation of 1) [9, 10]. The Kendall rank can be evaluated as in (3.14) and is denoted by the symbol τ:

$$\tau = \frac{2}{n(n-1)} \sum sign(x_i - x_j) \cdot sign(y_i - y_j) \tag{3.14}$$

where the operator $sign$ retrieve the sign of what is inside the parenthesis, and i, j indicates the data pairs.

The Spearman correlation index (3.15) is a measurement of the degree of relationship between two variables, with the hypothesis that they are predictable and continue [11]. In contrast to Pearson's linear correlation coefficients, Spearman's coefficients do not measure a linear relationship even when using interval measurements. In fact, this enables the stability of how well a relationship between two variables can be described using a monotonous function [12].

$$\rho = \frac{\sum_i (r_i - \hat{r})(s_i - \hat{s})}{\sqrt{\sum_i (r_i - \hat{r})^2} \cdot \sqrt{\sum_i (s_i - \hat{s})^2}} \tag{3.15}$$

where r and s represent the rank of the first and second variable respectively, while $\hat{r}$ and $\hat{s}$ is their average.

3.6 Analysis Results

In the following are shown the results, divided per biomass category, related to the data analysis performed in terms of regression, distribution and correlation indexes evaluation. As already stated, only the principal variables will be shown, and it is possible to find all the other figures and numerical values of this analysis in the database cloud.

3.7 Manure

Figure 3.3 and Table 3.3 summarize all the strongest correlations between the manure attributes using a linear model. As it is possible to see, not all the variables match a perfect line, due to the heterogeneity of the data, which further depends on its origin;

however, it is still possible to retrieve important information about specific variable trends and perform good estimations.

It is trivial that BMP and BD have a really similar trend, since if a biomass is highly biodegradable, then its methane potential is thereby high. But this is not always the case. In fact, it is also possible to have a highly biodegradable substrate,

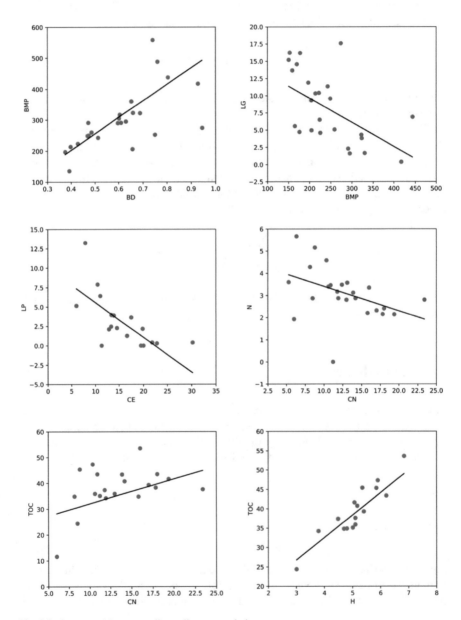

Fig. 3.3 Strongest Manure attributes linear correlations

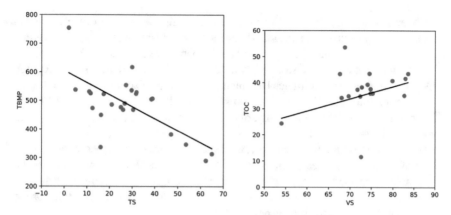

Fig. 3.3 (continued)

Table 3.3 Regression metrics for manure correlation analysis

Predictor	Regressor	R^2_{adj}	PCC	τ	ρ
BMP	BD	0.656	0.650	0.580	0.718
LG	BMP	0.664	− 0.584	− 0.493	− 0.035
N	CN	0.838	− 0.404	− 0.396	− 0.111
TOC	CN	0.920	0.427	0.233	0.343
TOC	H	0.835	0.915	0.762	0.898
TBMP	TS	0.955	− 0.291	− 0.083	− 0.113
TOC	VS	0.967	0.291	0.257	0.399

but the absolute CHONS composition does not allow to reach a good biogas purity. Therefore, adjustment with other biomass blending is necessary [13]. The R^2_{adj} of this regression is rather good for this kind of systems.

The trend of biomethane potential with respect to the lignin content is rather trivial. To have a higher biomethane value it is necessary to reduce the amount of lignocellulosic material [14, 15]. Lignin is not a degradable material and is considered as inert compound (i.e., not reacting species) when modelling [16]. The values of the correlation indexes become negative in this case since the lower the regressor, the higher the predictor.

Really interesting is the trend of the lipid content with respect to the cellulose. As it is possible to see, as the cellulose increase, the lipids decrease.

This is because both are macromolecule composition and together, with other variables as proteins, sugar and hemicellulose, define the total solid composition. It is important to not confuse with the volatile solid composition, which is made principally by sugar compounds, protein and lipids [17].

CN and N, and CN and TOC mathematical relation is straightforward, since is mathematically evaluated as the ratio between the total carbon and nitrogen content. As a confirmation to this, the regression performances reach really high scores.

TOC and H have a precise trend, with a really high correlation score. At high TOC content, correspond higher H content. This can be also translated in a higher content of cellulose/hemicellulose and sugar, which have a C/H ratio of 6/10 and 6/12 generally, respectively. Generally, manure-based substrates have a relatively high hydrogen content [18, 19].

Having a correlation between the TBMP and other variables proper of the composition of the biomass is really important. In fact, being the TBMP a numerical value obtainable from the elemental biomass composition, its relationship with the TS content can increase the precision when deriving the CHONS composition for the specific biomass from the biogas potential.

Finally, the TOC and VS relation is interesting for this particular biomass, since it indicates that by increasing the value of the organic carbon also VS increase, which in turns suggest the increase in the biogas production due to higher degradable content. Consequently, it is possible in this way to predict more precisely how much biogas will be produced.

It is now interesting to analyse the values obtained from the correlation indexes. Pearson coefficients indicate how much the model is linear or not. However, when its value approach to 0.5, it is possible to state that the data are too heterogeneous to establish the presence of linearity or non-linearity. Therefore, it is suggested to investigate other methods for the related variable.

TOC, H together with BMP, BD pairs have both a nice linearity score. The other pairs, despite the possibility to visualize and interpret their trend as linear, are not well scored. This is primarily due to the heterogeneity of the data.

Kendal τ helps to identify how much two variables are dependent between each other. The closer the value gets to 1, the higher the dependence is. Also, for this index, the higher score is achieved by the TOC, H and BMP, BD pairs, with a good score also reached by the LG, BMP and N, CN. The fact that LG, BMP score confirm the negative influence (inversely proportional) of the lignin on the methane production.

On the other hand, Spearman ρ establish how much a relation between two variables can be represented using a monotonous function. It is interesting at this point to evaluate the lowest-score pairs obtained. These reveals the possible presence of a minimum or maximum in these data relation, and so the possibility to find a value that optimize these attributes when in relation together. The lowest score is achieved by LG, BMP pair. In fact, it is not correct to say that the lignin is not degradable at all, since in determined circumstances, it is possible to have its biodegradation. However, when its content is too high, exceeding a specific amount, it starts acting like a poison for the degradation, reducing the biogas production [20]. Thus, this index suggests looking for a non-linear relation with the presence of a notable point as a minimum or maximum to better represent the real trends of these variables.

3.8 Agricultural Waste

Figure 3.4 and Table 3.4 summarize all the principal correlations between the attributes related to agricultural waste using a linear model.

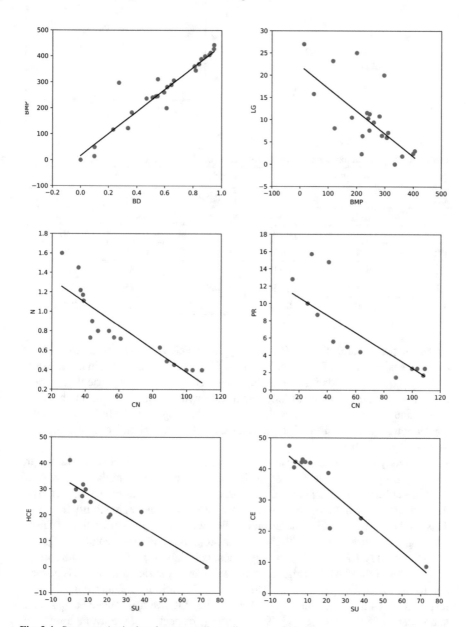

Fig. 3.4 Strongest Agricultural waste attributes linear correlations

Fig. 3.4 (continued)

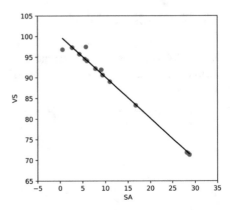

Table 3.4 Regression metrics for agricultural waste correlation analysis

Predictor	Regressor	R^2_{adj}	PCC	τ	ρ
BMP	BD	0.895	0.946	0.871	0.936
LG	BMP	0.324	− 0.686	− 0.481	− 0.658
N	CN	0.786	− 0.887	− 0.925	− 0.972
PR	CN	0.686	− 0.847	− 0.719	− 0.867
HCE	SU	0.817	− 0.907	− 0.687	− 0.858
CE	SU	0.856	− 0.927	− 0.718	− 0.844
VS	SA	0.986	− 0.993	− 0.905	− 0.961

As for the manure, the BMP is strictly correlated with the biodegradability of the biomass, especially for this category. In fact, while manure is primary made by carbohydrates and lipids that are highly degradable compounds, agricultural waste has a really high percentage of long-chain carbohydrates, as cellulose and hemicellulose, and lignin [21, 22, 23], which decrease the biogas productivity. The related correlation indexes confirm this trend and the fact that this relation can be interpreted as a linear correlation.

As already mentioned for the manure, lignin content and biomethane potential are inversely proportional one each other. It is interesting to see that the BMP derived from agricultural waste is not so lower with respect to the manure one. In fact, from the statistical point of view, both biomasses have a similar biodegradability (Fig. 3.5). This can be seen as an effect of the animal diet, which solely eat the yard and garden waste, and produce manure with the similar properties and lignin content [23].

In this case N, CN pair has a stronger linear correlation, confirmed by the correlation indexes (Table 3.4). The high CN values reached by agricultural waste is primary due to the lower nitrogen content. The average CN ratio for manure is 12.83, with respect to 57.04 for agricultural waste. Despite the TOC average content being similar and equal to 38.26%w/w$_{TS}$ and 39.75%w/w$_{TS}$; respectively, the average nitrogen content in manure is 3.04%, while for agricultural waste is 0.82%.

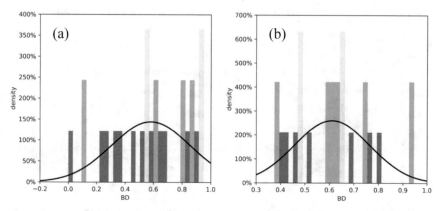

Fig. 3.5 Similarities between the biodegradability of the agricultural waste (**a**) and the manure (**b**)

The nitrogen amount is strictly influenced by the protein content, which is its primary source in terms of anaerobic digestion. Consequently, the decrease of protein content increases the CN value, as also shown in the figure. The lower Kendall value with respect to the N, CN pair however suggest investigating this relation with other higher order models.

The pairs HCE, SU and CE, SU are really similar one another, also in terms of score indexes, revealing a strong linear correlation. However, this is rather trivial, since the total carbohydrates content can be divided into the contribution of cellulose, sugars and hemicellulose. However, cellulose and hemicellulose have correlated one another since they both contribute to the plant cell structure, and this is the principal reason why both have a same trend with respect to the sugar.

Finally, the best score is achieved by the VS, SA pair, which has, as it is possible to see, a strong linear correlation. The content of salts in the biomass is usually related to the ash content discovered when biomass undergo to complete combustion [24, 25]. Due to their nature, salts are not biodegradable and are considered as an inert in the digestion processing. Consequently, it is reasonable to have an inversely proportional steep slope in this correlation, since the higher they are the lower the volatile solids content (Fig. 3.6).

3.9 Organic Waste

Figure 3.7 and Table 3.5 report the principal linear correlations between the attributes of organic waste category.

It is crucial to identify the principal correlations regarding the biodegradability and methane potential. Organic waste represents a really wide category, ranging from biomasses like fruit, vegetable, municipal organic waste fraction, to blood and slaughterhouse waste. As a consequence, it is much more difficult to establish a

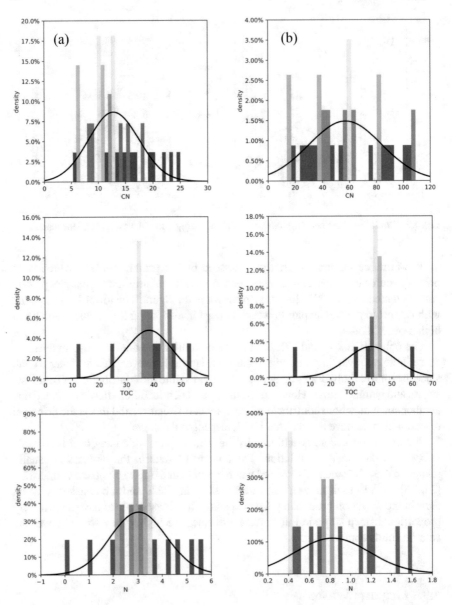

Fig. 3.6 Comparison between the CN ratio, TOC and N content in manure (**a**) and agricultural waste (**b**)

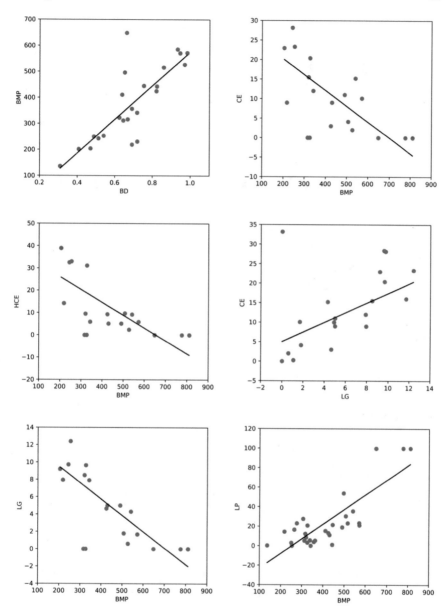

Fig. 3.7 Strongest Organic waste attributes linear correlations

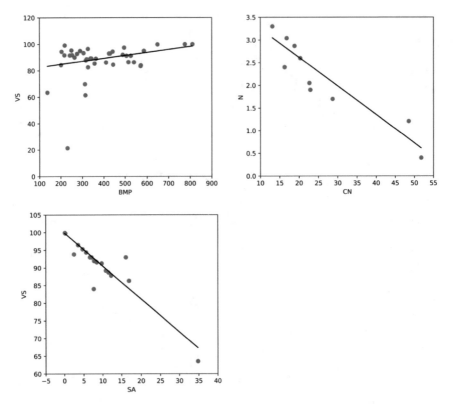

Fig. 3.7 (continued)

Table 3.5 Regression metrics for organic waste correlation analysis

Predictor	Regressor	R^2_{adj}	PCC	τ	ρ
BMP	BD	0.634	0.797	0.623	0.752
HCE	BMP	0.625	− 0.643	− 0.520	− 0.649
CE	BMP	0.665	− 0.615	− 0.428	− 0.540
LG	BMP	0.494	− 0.725	− 0.568	− 0.683
LP	BMP	0.652	0.808	0.485	0.653
VS	BMP	0.909	0.330	0.125	0.185
N	CN	0.840	− 0.925	− 0.867	− 0.927
TS	CN	0.164	0.431	0.379	0.549
CE	LG	0.420	0.674	0.668	0.758
VS	SA	0.849	− 0.922	− 0.789	− 0.867

correlation between biomasses of different nature. The average standard deviation of the attributes in this category is the highest with respect to the others. Thus, indexes values are meanly lower than other categories (Table 3.5), but still giving a good evaluation of the correlation wellness.

Also in this case, methane potential and biodegradability are strictly linked, and the same is found between the BMP and the lignin content LG.

Despite their heterogeneity, it is possible to highlight an inverse proportional relation between the BMP and both cellulose CE and hemicellulose HCE content, principally due to the presence in the organic waste category of fibrous biomasses as fruit and vegetables. This also justify the trend of the CE, LG pair, which highlight the natural relation between these two substances, that contributes to the building block of the plant cell's structure.

An interesting and strong correlation is between the lipid content LP and the BMP. The biodegradability of the lipids is really high [26]. Thus, the higher the concentration the higher the methane potential. Furthermore, as it is possible to see, there are three points at almost 100%w/w$_{TS}$ LP, which are data regarding the exhaust kitchen oil, which structurally is made by fats, reaching a really high level of methane potential, since lipids is the most biodegradable macromolecule together with sugar compounds.

Trivial is the relation between the VS and BMP, since volatile solids are the building block of the methane production through digestion. However, it is very interesting to notice that there exist biomasses that, also having a really high VS content, have a very low BMP value. This underlines the importance to also study the influences of other attributes for the success of the process. Despite the really high R^2_{adj}, Kendall and Spearman indexes suggest investigating for other mathematical model.

Also in this case, nitrogen and CN ratio are linearly correlated to one another, confirming the trend studied for the previous categories. As a confirmation to this, the values obtained by the correlation indexes are very high despite the lower number of data available.

Finally, also for the organic waste, volatile solids and salt content are strongly linearly correlated, probably due to the presence of some similar biomasses between these two categories.

3.10 Sludges

For this category, due to the lack of many data concerning the elemental composition, it is difficult to find robust correlations. Figure 3.8 reports the principal ones, with the relative metrics and correlation scores shown in Table 3.6.

Coherently with the previous biomasses studied, the pair BMP, BD has linked one another, as a confirmation to this correlation in every biomass.

It is interesting the influence of hydrogen on the sludge attributes: the increase in TOC leads to an increase in presence of lipids and carbohydrates. These substances

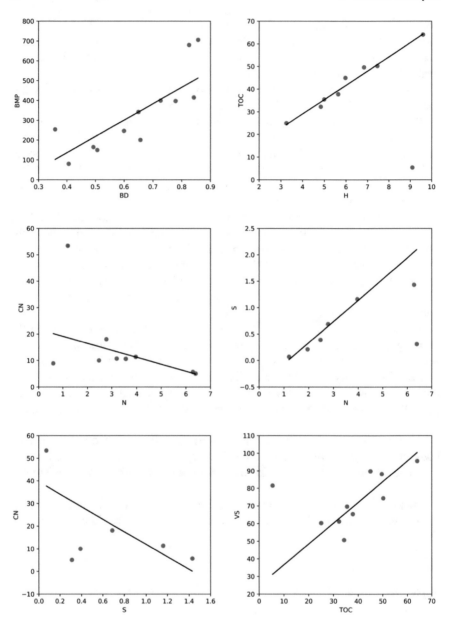

Fig. 3.8 Important Sludges attributes correlations

Table 3.6 Regression metrics for sludges correlation analysis

Predictor	Regressor	R^2_{adj}	PCC	τ	ρ
BMP	BD	0.673	0.824	0.697	0.823
TOC	H	0.975	0.229	0.611	0.519
CN	N	0.221	− 0.519	− 0.333	− 0.496
S	N	0.915	0.583	0.619	0.640
CN	S	0.280	− 0.558	− 0.200	− 0.252
VS	TOC	0.713	0.446	0.467	0.554

are very high in hydrogen, and in particular primary sludge, that already undergo to previous treatments and sedimentations stages before anaerobic digestion in wastewater treatment plants [27].

For this reason, it is interesting to see that the nitrogen content does significantly not affect the CN ratio as in the previous category. In fact, it is possible that other nitrogen-based compounds have been produced during the previous processes (i.e., ammonium) [28–30].

However, nitrogen content still influences the sulphur content in the sludge. The pair S, N in the figure shows a good correlation between these two quantities, coming both from the protein structure.

It is interesting to notice how the pair S, CN shows the possible presence of a maximum values in CN value when the sulphur is around 0.8%. This is in agreement with the literature [31], and highlights the possible presence of other substances rather protein that affect CN, N and S values. Of course, in this case is misleading referring to the correlation indexes.

Finally, the pair VS, TOC shows an interesting trend already seen in the previous category, through which the higher the volatile solids, the higher the organic carbon, which as consequence underline the good biodegradability of the sludge.

Example 3.3 Calculate the statistical index and the performances of the combination cellulose-sugar in the agricultural waste category.

Resolution

To perform all the calculations, here is the matlab code:

```
%% Federico Moretta (c)
%% Example 3.3
% Data (gather from directory where the database file is)
complete_db = readtable("./complete_database.csv");
% Select rows for only for desired category
category = complete_db(strcmp(complete_db.category, "agricultural_waste"),:);
CE = category.cellulose__TSw_;
SU = category.sugars__TSw_;
% Linear correlation evaluation
mdl = fitlm(CE, SU)
% Statistical indexes
pearson_corr = corr(CE, SU, 'Type', 'Pearson');
spearman_corr = corr(CE, SU, 'Type', 'Spearman');
kendall_corr = corr(CE, SU, 'Type', 'Kendall');
% Display the results
disp(['Pearson correlation coefficient_: ', num2str(pearson_corr)]);
disp(['Spearman correlation coefficient: ', num2str(spearman_corr)]);
disp(['Kendall correlation coefficient_: ', num2str(kendall_corr)]);
```

And this results in:

```
mdl =
Linear regression model:
y ~ 1 + x1
Estimated Coefficients:
Estimate SE tStat pValue
```

	Estimate	SE	tStat	pValue
(Intercept)	54.251	5.0838	10.671	5.1689e−15
x1	− 1.4289	0.19674	− 7.2629	1.3987e−09

```
Number of observations: 57, Error degrees of freedom: 55
Root Mean Squared Error: 13.6
R-squared: 0.49, Adjusted R-Squared: 0.48
F-statistic vs. constant model: 52.8, p-value = 1.4e−09
Pearson correlation coefficient_: − 0.69969
Spearman correlation coefficient: − 0.62588
Kendall correlation coefficient_: − 0.46739
>>
```

Do It Yourself

Calculate the statistical index of the combination cellulose-sugar in the organic waste category and compare it to the previous exercise.

Results

Pearson correlation coefficient: − 0.24074.

Spearman correlation coefficient: − 0.18981.

Kendall correlation coefficient: − 0.15419.

References

1. Quinn TP, Erb I, Richardson MF, Crowley TM (2018) Understanding sequencing data as compositions: an outlook and review. Bioinformatics 34(16):2870–2878. https://doi.org/10.1093/bioinformatics/bty175

2. Ross SM (2021) Chapter 6: distributions of sampling statistics. In: Ross SM (ed) Introduction to probability and statistics for engineers and scientists (6th edn). Academic Press, New York, 221–244. https://doi.org/10.1016/B978-0-12-824346-6.00015-6.

3. Yvonnet J, Monteiro E, He Q-C (2013) Computational homogenization method and reduced database model for hyperelastic heterogeneous structures. Int J Multiscale Comput Eng 11:3. https://doi.org/10.1615/IntJMultCompEng.2013005374

4. Kiernan D (2023) Chapter 7: correlation and simple linear regression. https://milnepublishing.geneseo.edu/natural-resources-biometrics/chapter/chapter-7-correlation-and-simple-linear-regression/

5. Benesty J, Chen J, Huang Y, Cohen I (2009) Pearson correlation coefficient. In: Noise reduction in speech processing, in Springer topics in signal processing, vol 2. Springer, Berlin, pp 1–4. https://doi.org/10.1007/978-3-642-00296-0_5

6. Farebrother RW (2017) Linear least squares computations. Routledge, New York. https://doi.org/10.1201/9780203748923

7. Agawin NSR, Duarte CM, Agustí S (2000) Nutrient and temperature control of the contribution of picoplankton to phytoplankton biomass and production. Limnol Oceanogr 45(3):591–600. https://doi.org/10.4319/lo.2000.45.3.0591

8. Tokuşoglu O, Unal MK (2003) Biomass nutrient profiles of three microalgae: spirulina platensis, chlorella vulgaris, and isochrisis galbana. J Food Sci 68(4):1144–1148. https://doi.org/10.1111/j.1365-2621.2003.tb09615.x

9. Michalik M, Wilczyńska-Michalik W (2012) Mineral and chemical composition of biomass ash. https://doi.org/10.13140/2.1.4298.5603

10. Gil A, Toledo M, Siles JA, Martín MA (2018) Multivariate analysis and biodegradability test to evaluate different organic wastes for biological treatments: anaerobic co-digestion and co-composting. Waste Manag 78:819–828. https://doi.org/10.1016/j.wasman.2018.06.052

11. Myers L, Sirois MJ (2006) Spearman correlation coefficients, differences between. In: Encyclopedia of statistical sciences. Wiley, Amsterdam. https://doi.org/10.1002/0471667196.ess5050.pub2

12. de Winter JCF, Gosling SD, Potter J (2016) Comparing the Pearson and Spearman correlation coefficients across distributions and sample sizes: a tutorial using simulations and empirical data. Psychol Methods 21(3):273–290. https://doi.org/10.1037/met0000079

13. Atkinson CF, Jones DD, Gauthier JJ (1996) Biodegradability and microbial activities during composting of poultry litter. Poult Sci 75(5):608–617. https://doi.org/10.3382/ps.0750608

14. Ahmadi-Pirlou M, Ebrahimi-Nik M, Khojastehpour M, Ebrahimi SH (2017) Mesophilic co-digestion of municipal solid waste and sewage sludge: effect of mixing ratio, total solids, and alkaline pretreatment. Int Biodeterior Biodegrad 125:97–104. https://doi.org/10.1016/j.ibiod.2017.09.004

15. Jain S, Jain S, Wolf IT, Lee J, Tong YW (2015) A comprehensive review on operating parameters and different pretreatment methodologies for anaerobic digestion of municipal solid waste. Renew Sustain Energy Rev 52:142–154. https://doi.org/10.1016/j.rser.2015.07.091

16. Batstone DJ et al (2002) The IWA anaerobic digestion model No 1 (ADM1). Water Sci Technol 45(10):65–73. https://doi.org/10.2166/wst.2002.0292

17. Liu J, Smith SR (2022) The link between organic matter composition and the biogas yield of full-scale sewage sludge anaerobic digestion. Water Sci Technol 85(5):1658–1672. https://doi.org/10.2166/wst.2022.058

18. Siddique MNI, Wahid ZA (2018) Achievements and perspectives of anaerobic co-digestion: a review. J Clean Prod 194(1):359–371. https://doi.org/10.1016/j.jclepro.2018.05.155

19. Wang X, Yang G, Feng Y, Ren G, Han X (2012) Optimizing feeding composition and carbon-nitrogen ratios for improved methane yield during anaerobic co-digestion of dairy, chicken manure and wheat straw. Bioresour Technol 120:78–83. https://doi.org/10.1016/j.biortech.2012.06.058

20. Benner R, Maccubbin AE, Hodson RE (1984) Anaerobic biodegradation of the lignin and polysaccharide components of lignocellulose and synthetic lignin by sediment microflora. Appl Environ Microbiol 47(5):998–1004

21. Wang M, Li W, Li P, Yan S, Zhang Y (2017) An alternative parameter to characterize biogas materials: available carbon-nitrogen ratio. Waste Manag 62:76–83. https://doi.org/10.1016/j.wasman.2017.02.025

22. Turner BL (2010) Variation in pH optima of hydrolytic enzyme activities in tropical rain forest soils. Appl Environ Microbiol 76(19):6485–6493. https://doi.org/10.1128/AEM.00560-10

23. Sutton A et al (2006) Manipulation of animal diets to affect manure production, composition and odors: state of the science. In: Animal agriculture and the environment, national center for manure and animal waste management white papers. ASABE, St. Joseph. https://doi.org/10.13031/2013.20259

24. Tang J, Wang XC, Hu Y, Zhang Y, Li Y (2017) Effect of pH on lactic acid production from acidogenic fermentation of food waste with different types of inocula. Bioresour Technol 224:544–552. https://doi.org/10.1016/j.biortech.2016.11.111

25. Cheah Y-K, Vidal-Antich C, Dosta J, Mata-Álvarez J (2019) Volatile fatty acid production from mesophilic acidogenic fermentation of organic fraction of municipal solid waste and food waste under acidic and alkaline pH. Environ Sci Pollut Res 26(35):35509–35522. https://doi.org/10.1007/s11356-019-05394-6

26. Li Y, Jin Y, Borrion A, Li H, Li J (2017) Effects of organic composition on the anaerobic biodegradability of food waste. Bioresour Technol 243:836–845. https://doi.org/10.1016/j.biortech.2017.07.028

27. Hanum F et al (2023) Treatment of sewage sludge using anaerobic digestion in Malaysia: current state and challenges. Front Energy Res 7:7. https://doi.org/10.3389/fenrg.2019.00019

28. Labatut RA, Pronto JL (2018) Chapter 4—sustainable waste-to-energy technologies: anaerobic digestion. In: Trabold TA, Babbitt CW (eds) Sustainable food waste-to-energy systems. Academic Press, New York, pp 47–67. https://doi.org/10.1016/B978-0-12-811157-4.00004-8

29. Li D et al (2015) Effects of feedstock ratio and organic loading rate on the anaerobic mesophilic co-digestion of rice straw and cow manure. Bioresour Technol 189:319–326. https://doi.org/10.1016/j.biortech.2015.04.033

30. Jabeen M, Zeshan S, Yousaf S, Haider MR, Malik RN (2015) High-solids anaerobic co-digestion of food waste and rice husk at different organic loading rates. Int Biodeterior Biodegrad 102:149–153. https://doi.org/10.1016/j.ibiod.2015.03.023

31. Dewil R, Baeyens J, Roels J, Steene BVD (2008) Distribution of Sulphur Compounds in Sewage Sludge Treatment. Environ Eng Sci 25(6):879–886. https://doi.org/10.1089/ees.2007.0143

Chapter 4
Synergisms and Antagonisms of Biomasses

Anaerobic Co-Digestion consists of the simultaneous digestion of two or more substrates. This technology has gained a lot of attention nowadays because it gives the possibility to significantly improve process performances [1, 2]. Indeed, the digestion of a single substrate might lead to poor outcomes in terms of substrate utilization and methane yield due to the lack of some nutrients or non-optimal parameters. By co-digesting different substrates together, that show "complementary" characteristics, instead, methane yield and process stability can be significantly improved, and synergistic effects may be observed too. On the other hand, an improper choice of co-substrates could lead to a system imbalance and create antagonistic effects, reducing the methane generation to mono-digestion [3, 4].

The feedstock for anaerobic digestion can be defined as any substrate—ranging from readily degradable to complex high-solids wastes—which can be turned into biogas and a digestate following the reactions described before.

Historically, anaerobic digestion has been associated with the treatment of animal manure and wastewater active sludge [5]; however, after the 70 s, the increasing demand for new waste management strategies and the necessity of renewable energy forms, broadened the field of applications for this technology and new substrates started to be exploited: for example, waste such as harvest remains, energy crops, garden wastes, algal biomasses, food wastes, municipal waste, industrial waste and wastewaters are currently used as possible feedstocks.

As previously mentioned, anaerobic co-digestion involves the use of mixtures of complementary substrates as feedstocks, which brings economic and process advantages since it allows for recovery of many typologies of waste coming from nearby sources, such as farms, industries, wastewater treatment plants [6].

Co-digestion can bring benefits such as an improvement in the management of wastewater through better capacity sparing [7, 8]; Improve in farming capabilities of the leftover biomass (i.e., digestate) after digestion, improving soil quality (i.e., pH, drainage, texture) and fertility [9–12]; increased biogas purity and biomethane yield [13]; better nutrient recovery as nitrogen and phosphorus, useful for fertilizer

© The Author(s), under exclusive license to Springer Nature Switzerland AG 2024
F. Moretta and G. Bozzano, *Mathematical and Statistical Approaches for Anaerobic Digestion Feedstock Optimization*, SpringerBriefs in Energy,
https://doi.org/10.1007/978-3-031-56460-4_4

production and farming activities [14]; increase the biomass resistance to malus agents, as antibiotics and chemical species as limonene, which both have an important inhibitory effects on bacteria metabolism [15, 16]; and considerably increase the process stability, in terms of biogas production and biomass activity [17].

However, to have a considerable increment of performances, the different substrates should be mixed in appropriate ratios so that the methane yield can be significantly improved to the mono-digestion, avoiding antagonistic effects too. In the work of Jacqueline et al. [18], 12 co-digestion blends between food waste and dairy manure were analysed to find the best loading ratio. A co-digestion performance index has been proposed, evaluated as the ratio between the biomethane potential of the proposed blend and the weighted average of the biomethane potential of the species composing the blend in terms of volatile solid content. It helped to recognize the presence of a synergistic effect of an antagonistic one. A range of -5% up to $+20\%$ oscillation in the BMP value has been obtained with different blend ratios.

Nielfa et al. [19] studied the optimum blending between the organic fraction of municipal solid waste (OFMSW) and the biological sludge (BS), revealing how powerful can the biomethane potential parameter be in its evaluation. They revealed that the blending ratio 80/20 OFMSW/BS has the highest BMP, equal to 220.6 mL_{CH_4}/g_{VS}, to the pure ones (201.5/164.5, respectively). Also, biomass of the same nature can undergo process performance variations when blended in different amounts. It is the case of Neves et al. (2006), which have processed five coffee wastes at different percentages together: traditional, barley, rye, malted barley and chicory. BMP assays revealed that the ratio 54/32/0/0/23, respectively, has the highest methane production, which exceeds 0.28 $m^3_{CH_4}/kg_{VS}$. It is interesting to see that, despite the optimum blending having 32% barley waste, the digestion of 100% barley brings to only 0.02 $m^3_{CH_4}/kg_{VS}$, thus revealing the presence of *blending* effects of biomasses of the same nature. The same considerations and results have been obtained also for biomasses based on organic waste, which is highly heterogenous and difficult to characterise uniquely, also in terms of nutrient and carbon balance [17, 20, 21]. Many other authors have addressed these points showing how crucial the contribution of co-digestion is toward the production of better-quality biogas. On the other hand, as seen previously, not all the biomasses can be mixed, and not always a maximum in the biomethane production during blending. It is possible to undergo a bad mixing ratio, revealing antagonistic effects between the biomasses considered [22]. In his work, he found that mixing in an equal amount of slurry and silage significantly inhibits methanogenic activity. Many research studies have demonstrated that some of the parameters characterizing the feedstock should be kept within defined ranges to obtain high methane yields, and mixtures of substrates should respect such constraints to have good performances. AcoD represents a beneficial technology also from this point of view since it allows the dilution of these compounds by other substrates, reducing their impact on methane yield. In the work of Poulsen and Adelard [23], which tests different mixes of cow dung and grass, an interesting trend is shown. When mixing cow dung and grass at a ratio of 0.67/0.33, the methane yield increases at 150 L kg^{-1}, but when switching to a ratio of 0.84/0.16, the methane yield drops down to 125 L kg^{-1}, which is much lower than the grass yield (226 L kg^{-1}) and

Fig. 4.1 Methane yield of co-digestion of cow dung and grass at different blending ratios (adapted from [22])

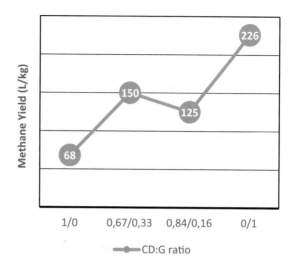

slightly higher than the cow dung (68 L kg^{-1}), revealing the presence of a slight antagonism. Figure 4.1 shows this interesting trend.

Xie et al. [24] have studied the presence of synergistic or antagonistic effects in microorganisms when co-digesting food waste and sewage sludge. They have shown that when the organic loading rate (OLR) of the unit is too high, methanogen inhibition occurs due to the VFA accumulation [25]; revealing that it is possible to find an optimal OLR for the configuration under study.

Finally, in the work of Montesdeoca-Pichucho et al. [26] it has been studied that the lag-time, which is the time that influences the starting of an enzymatic reaction, is highly affected when co-digesting more feedstock together; showing the possible presence of positive and negative impact with different blending ratios of different biomasses.

In the next section, the modelling of these synergistic and antagonistic effects will be addressed, revealing how it is possible to evaluate, for many feedstocks, the presence of an optimal blending ratio.

4.1 Biomethane Potential Analysis

The Bio-Methane Potential (BMP) is one of the principal indicators of the reactor performances. It indicates the maximum amount of methane that can be produced per grams of volatile solids entering the system [27]. A good prevision of this parameter, which is unique for every biomass, can be crucial to identify the feasibility of the process in terms of biogas production, costs evaluation and energy production. As stated from the previous chapters, the BMP can be evaluated numerically through the

Buswell formula, knowing the biomass elemental composition, or experimentally, with biomass specific batch assay.

Finding relations that correlate the BMP to biomass attributes can be crucial to find the optimal amount of biomass for a specific digestion or blending.

In this chapter, these relations are investigated, dividing the study by biomass category.

The BMPs of each category follow an interesting distribution (Fig. 4.2), which is very uniform for manure and agricultural waste, while having a higher deviation for the organic waste and sludges. This is due to the heterogenous nature of the biomasses considered in the database, as already explained in the chapter before.

In order to evaluate the BMP of a mixture, it is important to define no only its weighted sum with respect to the substrate mass composition (Eq. 4.1), but it is necessary to account for synergies between the biomasses [28].

$$BMP_{mix} = \sum_i x_i BMP_i \qquad (4.1)$$

In fact, biomasses of different nature interact with each other at microorganism level, leading as consequence to the increase (or decrease) of the biomethane productivity. This strictly depends on the type of biomass chosen for the blending, since their attributes can drastically influence the fermentative process.

To account for this behavior, the equation that is proposed is the following [29]:

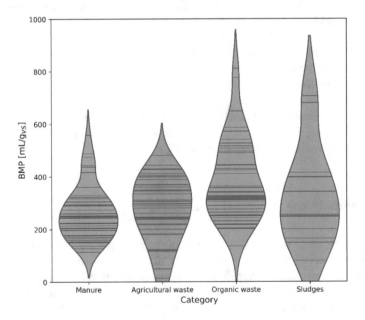

Fig. 4.2 Violin plot of BMP for every biomass category

$$BMP_{max} = \sum_i x_i BMP_i + \sum_{i \neq j} x_i x_j BMP_{acod} \qquad (4.2)$$

This equation calculates the biomethane potential for a blending of two substrates, which is the most typical one. The term BMP_{acod} is the interaction term and must be evaluated looking forward the attribute of every biomass involved. This is calculated from the parameters that best resemble the evaluation of the process performances and stability. Between all the parameters analyzed in the previous chapter, the biodegradability and the carbon-to-nitrogen ratio have been selected. In fact, while biomethane potential is directly proportional to the biodegradability, the C/N suggest the stability of the blending chosen. A substrate with a high carbon content or a lower nitrogen content, and consequently a high C/N value biomass should be mixed with another one that has lower C/N value. This can balance this property into the optimal range. Of course, the biodegradability should be the highest one possible, but this should always be coherent with the C/N value obtained. Thus, the interaction term can be written as follow:

$$BMP_{acod} = \beta_0 + \beta_1 \cdot \left(\frac{C}{N}\right)_{mix} + \beta_2 \cdot BD_{mix} + \beta_3 \cdot \left(\frac{C}{N}\right)_{mix}^2 + \beta_4 \cdot BD_{mix}^2$$
$$(4.3)$$

where the mixing parameters C/N_{mix} and BD_{mix} are evaluated as Eq. 4.1, while parameters β are regressed from the database (Table 4.1).

When more than two substrates are considered, Eq. 4.2 should consider the presence of a third factor, which comprise influence of other biomasses blended together (Eq. 4.4). Of course, the higher is the number of substrates blended, the more difficult would be to find a good compromise between the parameters calculating the BMP for a good and stale process. Thus, a good choice of the right feedstocks to mix is necessary a priori to these calculations.

$$BMP_{max} = \sum_i x_i BMP_i + \left(\sum_{i \neq j} x_i x_j + \prod_i x_i\right) BMP_{acod} \qquad (4.4)$$

Despite the simplicity of this expression, it is not the best way to solve a BMP maximization problem. In fact, with three or four substrates, the extended equation X.4 becomes respectively:

$$BMP_{max} = x_1 BMP_1 + x_2 BMP_2 + x_3 BMP_3$$

Table 4.1 BMPacod parameters values obtained through a regression from the database

β_0	β_1	β_2	β_3	β_4
21.66	1.256	445.7	− 0.022	− 7.820

$$+ (x_1x_2 + x_1x_3 + x_2x_3 + x_1x_2x_3)BMP_{acod} \tag{4.5}$$

$$BMP_{max} = x_1BMP_1 + x_2BMP_2 + x_3BMP_3 + x_4BMP_4$$
$$+ (x_1x_2x_3 + x_1x_2x_4 + x_1x_3x_4 + x_2x_3x_4 + x_1x_2x_3x_4)BMP_{acod} \tag{4.6}$$

As it is possible to observe, the term inside the parenthesis become smaller with the increase in the number of substrates, drastically reducing the relative effect of the BMP_{acod} term to the maximization problem. In Table 4.2 is shown an example of the effect aforementioned.

As it is possible to see from the table, with two biomasses, the global interaction term (x_1x_2) results to be 0.12, while having three biomasses blended this value rise up to 0.37 $(x_1x_2 + x_2x_3 + x_1x_3 + x_1x_2x_3)$. Then, when reaching a 4 or 5 feedstock mixture, the values drastically drop to 0.047 and 0.005, respectively. Figure 4.3 shows the impact of these interaction terms on the BMP_{acod} calculation, considering as maximum interaction possible the pure biomass.

At this point a three-biomass blending better register the effect of the mixing parameters as BD and C/N composing the BMP_{acod}, however, from the computational point of view, it is difficult for a solver to find a solution in a surface, and specific software's shall be used. To solve this issue, it is possible to calculate the global BMP of a + 2 blending through the introduction of a pseudo-substrate. Basically, a BMP maximization problem is solved through two of the biomasses selected. The solution represents a pseudo-substrate that has as properties the C/N_{mix} and BD_{mix} of

Table 4.2 Values of the interaction terms in Eqs. 4.4, 4.5 and 4.6. Blending composition value has been assumed for this study

Composition	Value	First term	Value	Second term	Value	Third term	Value
Case: 3 biomasses blending							
Substrate 1	0.3000	x_1	0.3000	x_1x_2	0.1200	$x_1x_2x_3$	0.0360
Substrate 2	0.4000	x_2	0.4000	x_2x_3	0.1200		
Substrate 3	0.3000	x_3	0.3000	x_1x_3	0.0900		
Case: 4 biomasses blending							
Substrate 1	0.3000	x_1	0.3000	$x_1x_2x_3$	0.0203	$x_1x_2x_3x_4$	0.0020
Substrate 2	0.4500	x_2	0.4500	$x_2x_3x_4$	0.0068		
Substrate 3	0.1500	x_3	0.1500	$x_1x_3x_4$	0.0045		
Substrate 4	0.1000	x_4	0.1000	$x_1x_2x_4$	0.0135		
Case: 5 biomasses blending							
Substrate 1	0.1500	x_1	0.1500	$x_1x_2x_3x_4$	0.0016	$x_1x_2x_3x_4x_5$	0.0002
Substrate 2	0.3500	x_2	0.3500	$x_1x_2x_3x_5$	0.0005		
Substrate 3	0.1000	x_3	0.1000	$x_1x_3x_4x_5$	0.0005		
Substrate 4	0.3000	x_4	0.3000	$x_1x_2x_4x_5$	0.0016		
Substrate 5	0.1000	x_5	0.1000	$x_2x_3x_4x_5$	0.0011		

Fig. 4.3 Interaction term
impact on the BMP
calculation for different
amount of biomasses
blended

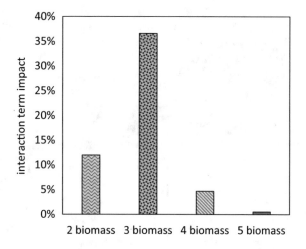

the optimal $x_1 - x_2$ calculated. Then, this pseudo-substrate (x^p) is then blended with another biomass on the list and the procedure is repeated until all the feedstocks have been used. This guaranteed stability on the solution seeking procedure and is fast to compute, since only a 2D objective function is considered. Figure 4.4 resemble the methodology just mentioned in a graphical scheme.

Example 4.1 Calculate the optimal blending ratio and the highest BMP achievable for the mixture; dairy manure—slaughterhouse residue—yard waste.

Resolution

Following the procedure explained in the previous chapter and showed in the Fig. 4.4, firstly, the optimal conditions for dairy manure and slaughterhouse residue are found. Data from PAD have been taken. A 50% blend is assumed.

Starting from the mixing properties, with the starting blend the result is:

$$CN_{mix} = \sum_i x_i CN_i = 17.5$$
$$BD_{mix} = \sum_i x_i BD_i = 0.65 \qquad BMP_{mix} = \sum_i x_i BMP_i = 294.98 \text{ mL}_{CH_4}/g_{VS}$$

Thus, the BMP_{acod} is computed as (4.3), giving as results 323.52 mL$_{CH_4}$/g$_{VS}$. Finally, the BMP_{max} of the blend is evaluated as (4.4). To maximize, the Solver Add-on in Microsoft Excel© has been used. The resulting BMP_{max} has a value of 380.18 mL$_{CH_4}$/g$_{VS}$, with a blending of 0.385/0.615 for dairy and slaughter respectively. The new mixing parameters become:

$$CN_{mix} = \sum_i x_i CN_i = 16.8$$
$$BD_{mix} = \sum_i x_i BD_i = 0.66 \qquad BMP_{mix} = \sum_i x_i BMP_i = 302.6 \text{ mL}_{CH_4}/g_{VS}$$

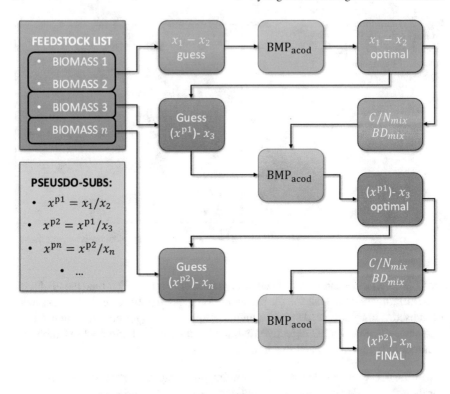

Fig. 4.4 Scheme of the BMP maximization through the pseudo-substrate routine

These now become the properties of the new pseudo-component x^p. Restarting with a new blend between the pseudo-component and the yard waste, assuming also in this case 50% ratio, the new mixing properties become:

$$CN_{mix} = \sum_i x_i CN_i = 32.0$$
$$BD_{mix} = \sum_i x_i BD_i = 0.54 \qquad BMP_{mix} = \sum_i x_i BMP_i = 233.04 \ \text{mL}_{CH_4}/g_{VS}$$

The new BMP_{acod} is computed as (4.3) gives as results 277.55 mL$_{CH_4}$/g$_{VS}$. Solving as before through a maximization, the new and final BMP_{max} become: 324.81 mL$_{CH_4}$/g$_{VS}$, which is right, because yard waste drops the biodegradability of the blend by adjusting the carbon to nitrogen ratio:

$$CN_{mix} = \sum_i x_i CN_i = 25.15$$
$$BD_{mix} = \sum_i x_i BD_i = 0.60 \qquad BMP_{mix} = \sum_i x_i BMP_i = 268.97 \ \text{mL}_{CH_4}/g_{VS}$$

Do It Yourself

Find the optimum blending ration between chicken manure, sludge waste and straw waste (use swage waste and straw waste as first pseudo component, do the switch one of these two substrates with chicken manure and compare the results). Compare the results with the previous exercise.

Results

Chicken manure fraction	0.38
Sewage sludge fraction	0.20
Straw waste	0.42
BMP_{max}	278.76 mL_{CH_4}/g_{VS}

References

1. Croce S, Wei Q, D'Imporzano G, Dong R, Adani F (2016) Anaerobic digestion of straw and corn stover: the effect of biological process optimization and pre-treatment on total bio-methane yield and energy performance. Biotechnol Adv 34(8):1289–1304. https://doi.org/10.1016/j.biotechadv.2016.09.004

2. Xie S et al (2016) Anaerobic co-digestion: a critical review of mathematical modelling for performance optimization. Bioresour Technol 222:498–512. https://doi.org/10.1016/j.biortech.2016.10.015

3. Jain S, Jain S, Wolf IT, Lee J, Tong YW (2015) A comprehensive review on operating parameters and different pretreatment methodologies for anaerobic digestion of municipal solid waste. Renew Sustain Energy Rev 52:142–154. https://doi.org/10.1016/j.rser.2015.07.091

4. Siddique MNI, Wahid ZA (2018) Achievements and perspectives of anaerobic co-digestion: A review. J Clean Prod 194(1):359–371. https://doi.org/10.1016/j.jclepro.2018.05.155

5. Abbasi T, Tauseef SM, Abbasi SA (2012) A brief history of anaerobic digestion and "biogas." In: Abbasi T, Tauseef SM, Abbasi SA (eds) Biogas energy, SpringerBriefs in environmental science. Springer, New York, NY, pp 11–23

6. Karki R et al (2021) Anaerobic co-digestion: current status and perspectives. Bioresour Technol 330:125001. https://doi.org/10.1016/j.biortech.2021.125001

7. Jones CA (2020) Food waste co-digestion at water resource recovery facilities. Consultato: 18 settembre 2023. Disponibile su: https://www.accesswater.org/publications/-10033428/food-waste-co-digestion-at-water-resource-recovery-facilities

8. Xie S, Higgins MJ, Bustamante H, Galway B, Nghiem LD (2018) Current status and perspectives on anaerobic co-digestion and associated downstream processes. Environ Sci Water Res Technol 4(11):1759–1770. https://doi.org/10.1039/C8EW00356D

9. Alburquerque JA et al (2012) Agricultural use of digestate for horticultural crop production and improvement of soil properties. Eur J Agron 43:119–128. https://doi.org/10.1016/j.eja.2012.06.001

10. Bustamante MA et al (2012) Co-composting of the solid fraction of anaerobic digestates, to obtain added-value materials for use in agriculture. Biomass Bioenergy 43:26–35. https://doi.org/10.1016/j.biombioe.2012.04.010

11. Rékási M et al (2019) Comparing the agrochemical properties of compost and vermicomposts produced from municipal sewage sludge digestate. Bioresour Technol 291:121861. https://doi.org/10.1016/j.biortech.2019.121861

12. Surendra KC, Tomberlin JK, van Huis A, Cammack JA, Heckmann L-HL, Khanal SK (2020) Rethinking organic wastes bioconversion: evaluating the potential of the black soldier fly (*Hermetia illucens* (L.)) (Diptera: Stratiomyidae) (BSF). Waste Manag 117:58–80. https://doi.org/10.1016/j.wasman.2020.07.050

13. Elsayed M et al (2020) Innovative integrated approach of biofuel production from agricultural wastes by anaerobic digestion and black soldier fly larvae. J Clean Prod 263:121495. https://doi.org/10.1016/j.jclepro.2020.121495

14. O'Brien BJ, Milligan E, Carver J, Roy ED (2019) Integrating anaerobic co-digestion of dairy manure and food waste with cultivation of edible mushrooms for nutrient recovery. Bioresour Technol 285:121312. https://doi.org/10.1016/j.biortech.2019.121312

15. Ruiz B, Flotats X (2016) Effect of limonene on batch anaerobic digestion of citrus peel waste. Biochem Eng J 109:9–18. https://doi.org/10.1016/j.bej.2015.12.011

16. Zhang L et al (2018) Fate of antibiotic resistance genes and mobile genetic elements during anaerobic co-digestion of Chinese medicinal herbal residues and swine manure. Bioresour Technol 250:799–805. https://doi.org/10.1016/j.biortech.2017.10.100

17. Astals S, Nolla-Ardèvol V, Mata-Alvarez J (2012) Anaerobic co-digestion of pig manure and crude glycerol at mesophilic conditions: biogas and digestate. Bioresour Technol 110:63–70. https://doi.org/10.1016/j.biortech.2012.01.080

18. Jacqueline H, Ebner Rodrigo A, Labatut Jeffrey S, Lodge Anahita A, Williamson Thomas A, Trabold (2016) Anaerobic co-digestion of commercial food waste and dairy manure: Characterizing biochemical parameters and synergistic effects. Waste Manag 52286-52294. https://doi.org/10.1016/j.wasman.2016.03.046

19. Nielfa A, Cano R, Fdz-Polanco M (2015) Theoretical methane production generated by the co-digestion of organic fraction municipal solid waste and biological sludge. Biotechnol Rep 514–521. https://doi.org/10.1016/j.btre.2014.10.005

20. Cabbai V, Ballico M, Aneggi E, Goi D (2013) BMP tests of source selected OFMSW to evaluate anaerobic codigestion with sewage sludge. Waste Manag 33(7):1626–1632. https://doi.org/10.1016/j.wasman.2013.03.020

21. Esposito G, Frunzo L, Giordano A, Liotta F, Panico A, Pirozzi F (2012) Anaerobic co-digestion of organic wastes. Rev Environ Sci Biotechnol 11(4):325–341. https://doi.org/10.1007/s11157-012-9277-8

22. Himanshu H, Murphy JD, Grant J, O'Kiely P (2018) Antagonistic effects on biogas and methane output when co-digesting cattle and pig slurries with grass silage in in vitro batch anaerobic digestion. Biomass Bioenergy 109:190–198. https://doi.org/10.1016/j.biombioe.2017.12.027

23. Poulsen TG, Adelard L (2016) Improving biogas quality and methane yield via co-digestion of agricultural and urban biomass wastes. Waste Manag 54118–54125. https://doi.org/10.1016/j.wasman.2016.05.020

24. Xie S, Wickham R, Nghiem LD (2017) Synergistic effect from anaerobic co-digestion of sewage sludge and organic wastes. International Biodeterioration & Biodegradation 116:191–197. https://doi.org/10.1016/j.ibiod.2016.10.037

25. Silvestre G, Illa J, Fernández B, Bonmatí A (2014) Thermophilic anaerobic co-digestion of sewage sludge with grease waste: effect of long chain fatty acids in the methane yield and its dewatering properties. Appl Energy 117:87–94. https://doi.org/10.1016/j.apenergy.2013.11.075

26. Montesdeoca-Pichucho NB, Garibaldi-Alcívar K, Baquerizo-Crespo RJ, Gómez-Salcedo Y. Pérez-Ones O, Pereda-Reyes I (2020) Synergistic and antagonistic effects in anaerobic co-digestion. Analysis of the methane yield kinetics Revista Facultad de Ingeniería Universidad de Antioquia. https://doi.org/10.17533/udea.redin.20220473

27. Angelidaki I et al (2009) Defining the biomethane potential (BMP) of solid organic wastes and energy crops: a proposed protocol for batch assays. Water Sci Technol J Int Assoc Water Pollut Res 59(5):927–934. https://doi.org/10.2166/wst.2009.040

28. Ferdeş M, Paraschiv G, Ionescu M, Dincă MN, Moiceanu G, Zăbavă BŞ (2023) Anaerobic co-digestion: a way to potentiate the synergistic effect of multiple substrates and microbial diversity. Energies 16(5):5. https://doi.org/10.3390/en16052116

29. Moretta F, Goracci A, Manenti F, Bozzano G (2022) Data-driven model for feedstock blending optimization of anaerobic co-digestion by BMP maximization. J Clean Prod 375:134140. https://doi.org/10.1016/j.jclepro.2022.134140

Chapter 5
Stability Parameters

The FOS/TAC parameter is used to monitor process stability by focusing on the pH value. FOS/TAC is defined as the ratio of the amount of volatile fatty acids that accumulate during anaerobic digestion to the alkalinity present in the system [1]. Since volatile fatty acids accumulate during the acidogenesis phase, they can lead to a consumption of alkalinity and consequently a decrease in pH. For this reason, it is of fundamental importance to have a buffer solution that can counteract the pH decrease bringing it back to neutral values, as the optimal pH value for an anaerobic digester is around 7 [2].

From the literature, it is found that the optimal range for FOS/TAC to maintain stable operation of the anaerobic digester is between 0.3 and 0.4.

In fact, for values > 0.4, the buffer solution's alkalinity is unable to maintain the system's pH on neutrality due to excessive production of volatile fatty acids; for values lower than 0.3, the digester can support a higher organic volumetric load.

Alkalinity (TAC) is mainly characterized by the coexistence of ammonia, generated by protein degradation, and the dissolution of CO_2 in the liquid promoting the formation of bicarbonate. The resulting buffer system is called the calcium-acetate system leading to the formation of sodium bicarbonate [3], which is a salt whose dissolution ensures high alkalinity, thus counteracting the decrease in pH caused by the accumulation of volatile fatty acids:

$$CO_2 + H_2O \leftrightarrow HCO_3^- + H^+ \tag{5.1}$$

$$HCO_3^- + NH_4^+ \leftrightarrow NH_4HCO_3 \tag{5.2}$$

Therefore, in the study carried out in the literature, to obtain an accurate estimate of alkalinity, the total amount of nitrogen obtained from proteins and the amount of CO_2 dissolved in the medium are evaluated for each substrate.

F. Moretta and G. Bozzano, *Mathematical and Statistical Approaches for Anaerobic Digestion Feedstock Optimization*, SpringerBriefs in Energy, https://doi.org/10.1007/978-3-031-56460-4_5

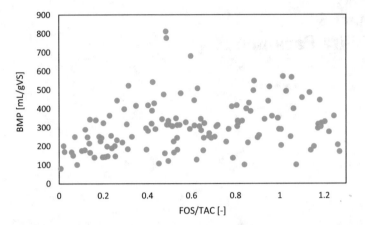

Fig. 5.1 FOS&TAC calculation results with respect to BMP

Regarding FOS, the literature presents a problematic issue as this parameter considers the acids in the system as if they were only acetic acid, neglecting all other types of acids formed during the acidogenesis phase [4].

Therefore, to obtain more accurate results, yield coefficients of the biomasses concerning these substances, taken from the literature [5, 6], were used in this study allowing a simpler but effective estimation of acetic acid, butyric acid, propionic acid and valeric acid production from the proteins and sugars in each substrate, which are considered as the main sources of system acidification. Only proteins and sugars are considered since no significant contribution has been shown from lipids degradation to FOS components.

Once the FOS/TAC value has been calculated, the biomethane yield trend to this stability parameter can be derived. An example can result in the trend shown in Fig. 5.1

From the graphs, it can be observed that the biomethane yield increases until reaching a maximum around the FOS/TAC value of 0.5, then it starts to decrease and stabilize within the range of 0.8–0.9.

Two key points should be emphasized from these results (Fig. 5.2):

1. The obtained results deviate slightly from what was expected from the literature due to more precise results thanks to the previously described coefficients (Table 5.1);
2. Beyond FOS/TAC values of 0.8–0.9, this parameter has a less significant influence on the biomethane yield. Therefore, subsequent studies will focus on the highlighted area of the graph.

At this point, focusing on the area of interest, the goal is to demonstrate the power of FOS/TAC to evaluate the stability of anaerobic digestion. Consequently, the two-parameter equation (see previous chapter) has been used to estimate the biomethane yield for a mixture of two substrates: Chicken Manure (CM) as referenced biomasses,

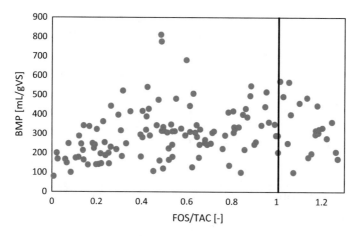

Fig. 5.2 FOS&TAC results with the zone of interest

Table 5.1 Yield coefficients used to calculate FOS values from macromolecule concentration

Parameter	Value	Macromolecule of origin	Final compound
Acetic acid yield from amino acids	0.40	Proteins	Acetic acid
Butyric acid yield from amino acids	0.26	Proteins	Butyric acid
Propionic acid yield from amino acids	0.05	Proteins	Propionic acid
Valeric acid yield from amino acids	0.23	Proteins	Valeric acid
Acetic acid yield from sugars	0.41	Carbohydrates	Acetic acid
Butyric acid yield from sugars	0.13	Carbohydrates	Butyric acid
Propionic acid yield from sugars	0.27	Carbohydrates	Propionic acid

and Sugar-Beet Byproducts (SBB) and Exhaust Kitchen Oil (EKO) as blending agent. The possible synergistic effects between them are taken into consideration. Equations used are (Eqs. 5.3–5.6):

$$BMP_{acod} = x_1 BMP_1 + x_2 BMP_2 + x_1 x_2 BMP_{mix} \tag{5.3}$$

$$BMP_{mix} = \beta_0 + \beta_1 \left(\frac{C}{N}\right)_{mix} + \beta_2 BD_{mix} + \beta_3 \left(\frac{C}{N}\right)_{mix}^2 + \beta_4 BD_{mix}^2 \tag{5.4}$$

$$\left(\frac{C}{N}\right)_{mix} = \sum_{i=i}^{NC} x_i \left(\frac{C}{N}\right) \tag{5.5}$$

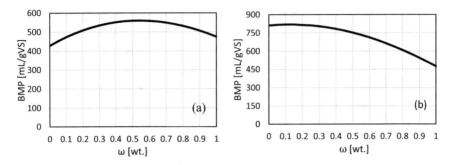

Fig. 5.3 **a** synergies found between CM (475 mL/g$_{VS}$) and SBB (428 mL/g$_{VS}$); and **b** antagonisms/absence of synergies found between CM (475 mL/g$_{VS}$) and EKO (811 mL/g$_{VS}$)

$$BD_{mix} = \sum_{i=1}^{NC} x_i BD_i \tag{5.6}$$

The β_i parameters are interaction coefficients obtained with a multidimensional regression analysis on all the biomasses, as described in the previous section.

To assess whether FOS/TAC can be considered a stability parameter, a reference substrate (represented by the red dot in Fig. 5.4) has been selected, and using the BMP_{mix} equation, synergy with all other substrates in the area of interest was evaluated one by one.

To determine if synergy was present between these substrates, the biomethane yield for each pair was plotted against the mixture composition. The presence of a maximum point between the two pure components is the signal of synergy between these biomasses.

However, if the trend resembled a monotony curve, no synergy is present at all. The results of both cases are shown in Fig. 5.3.

If observed from a C/N point of view it is possible to note that the obtained results are coherent with the acceptable values of C/N for an anaerobic digester; in fact, in the case of synergism we have a co-digestion between a substrate that is characterized by a high N content (CM, with N = 4.37%mol) and a substrate with a high C content (SBB, with C = 41.62%mol) leading to an optimal C/N, while in the case of antagonism we have a co-digestion between two substrates that are both rich in terms of C; CM has a C content of 40.16%mol, while EKO has a C content of 73.52%mol. So, this combination leads to a high value of C/N, leading to possible process instability.

The results of the study are shown in Fig. 5.4, where the reference substrate is highlighted in red. Substrates exhibiting a BMP peak value higher than 25% of the maximum BMP value in the considered blend are categorized as blend-in-synergy and are marked in green. On the contrary, blends with lower BMP peak values or those lacking this characteristic are categorized as not-in-synergy, and thus are marked in yellow.

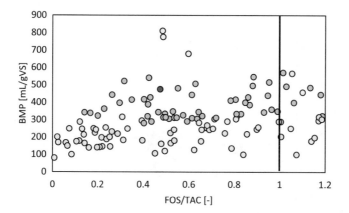

Fig. 5.4 FOS/TAC results for CM with highlighted regions of synergies (green dots) and regions of no synergies (yellow dots). In dark green, is indicated the SBB biomass, while in orange the EKO biomass

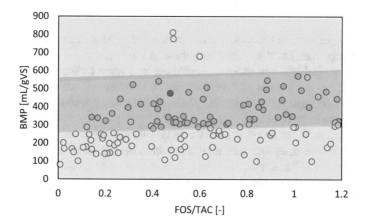

Fig. 5.5 Iso-regions detection in FOS/TAC analysis for CM

It can be observed that there are different regions where the closer a substrate is to the reference substrate, the higher the probability of synergy. Moreover, to validate the synergies presence, there are indicated in dark green and orange the two biomasses that show synergies and antagonism studied before and reported in Fig. 5.5, namely sugar-beet byproducts and exhaust kitchen oil, respectively.

As a confirmation, a second biomass has been analyzed, Sow Manure (SM), while the procedure for evaluating synergistic effects is the same as before. For this new case study, the trends obtained are shown in Fig. 5.6.

It is possible to observe that even under these new conditions, there are areas where it is possible to have a greater presence of conjugated biomass (i.e., biomasses that are in synergy with the referenced one) and thus a greater probability of having

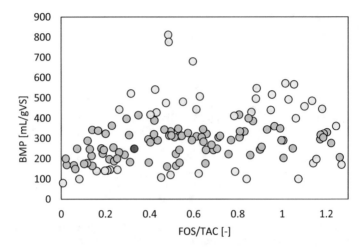

Fig. 5.6 FOS/TAC results for SM with highlighted regions of synergies (green dots) and regions of no synergies (yellow dots)

a stable and performant process, and this occurs for regions closer and closer to the reference substrate. On the other hand, there are also areas where the probability of no synergy is high, and these emerge more and more as we move away from the red dot. So, iso-regions of synergies can be highlighted, where biomasses can be classified by their level of process stability and quality when blended with the referenced one in an optimal solution.

In conclusion, it can be argued that the FOS/TAC parameter can be used to assess the stability of the anaerobic digester if it is within the influence zone, i.e., for FOS/TAC values < 0.8, but at the same time, it is important to consider two substrates that have sufficiently similar predicted FOS/TAC values to each other to achieve synergistic effects, thus being within the same iso-regions (Fig. 5.7).

In conclusion, it is possible to see that iso-regions are different for every biomass considered. However, by taking substrates that are near to the referenced one, there will be strong synergies effects. The biomasses that are in the green iso-regions are those that explicate this behavior and are called conjugated biomass. The ones in the yellow iso-regions do not show any synergies at all and won't influence significantly the digestion process but decrease the biomethane potential.

5.1 The Organic Loading Rate

The organic loading rate (OLR) is a very important parameter to define the operating conditions that characterize the process of anaerobic digestion; in fact, OLR represents the amount of volatile solids that must be fed to the reactor every day [7]. The volatile solids are the portion of organic matter inside the substrate while the

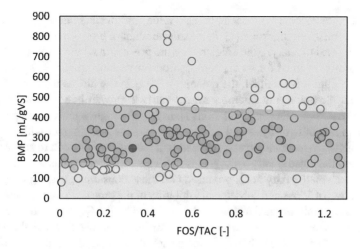

Fig. 5.7 Iso-regions detection in FOS/TAC analysis for SM

remaining part of solids is defined as fixed solids [7]. The amount of undegraded solids is given by the sum of fixed solids and a specific fraction of volatile solids since the substrate's biodegradability is never equal to one. OLR can be calculated through the following equation:

$$OLR = VS \cdot Q/V \qquad (5.7)$$

where VS [kg/m^3] is the concentration of volatile solids in the feeding; Q [m^3/day] is the inlet flow rate to the anaerobic digester, and V [m^3] is the volume of the reactor [8]. So OLR is obtained in terms of [kg$_{VS}$/m^3/day], but since the ratio between V and Q represents the hydraulic retention time [day], OLR can be also expressed as:

$$OLR = VS/HRT \qquad (5.8)$$

What is found in the literature is that by increasing the value of OLR, it is possible to reach a certain point in which the value of organic loading rate is so high that causes the inhibition of the process. It is very important to point out that the OLR value that induces inhibition changes from case to case, as a matter of fact it depends on several factors such as operating conditions and reactor configuration. Despite these numerous parameters that play an important role in the stability of the process, it was possible to note that in many papers the inhibition occurs for an OLR value equal to 6 [kg$_{VS}$/m^3/day] [9–12]. For this reason, it makes sense to consider 6 [kg$_{VS}$/m^3/day] as that organic loading rate value beyond which anaerobic digestion becomes unstable.

A very important aspect associated with OLR is that the negative effects caused by the reaching of its critical value can be observed in many useful parameters for defining the performance of the process, like the reduction in VS degraded percentage, that occurs for very high OLR value, and the amount of CO_2 produced during the anaerobic digestion [7]. In fact, the amount of biogas in terms of CO_2 and methane is used to analyze the efficiency of the process in which stable and optimal conditions are obtained for a CH_4 composition between 60 and 65% [7]. Once critical OLR values are obtained, the quantity of CO_2 in the biogas increases and this means that the acidifying microorganisms starts to prevail over the methanogens leading to a strong VFAs accumulation [7]. In fact, for high organic loading rate, the production rate of hydrolysis and acidogenesis steps is much higher with respect to the one related to methanogenesis step and so it results very difficult to consume in a correct way the volatile fatty acids [9, 11, 12]. This inefficiency causes an imbalance between the volatile fatty acids produced and those consumed, leading to a critical accumulation of VFAs and subsequent inhibition of the process. The amount of these two families of microorganisms is strongly affected by the OLR, but not only for its high values as described before, but also in start-up conditions in which the value of organic loading rate is low. In fact, in the presence of low OLR, methanogens are in a starving condition, in which their required nutrients are much higher than those fed [12]. Since it is not possible to satisfy in an appropriate way the methanogens diet, this group of microorganisms tends to die thus preventing the production of biomethane; so, it is necessary to take advantage of a higher OLR to keep methanogens alive.

Returning to the case study of a high OLR, it is important to highlight that a critical level of accumulated VFAs is reached, but if this problem is seen under a microorganism's point of view, why does process inhibition occur? The correct answer is related to the optimal pH range for methanogens growth [12]; In fact, a strong VFAs accumulation causes a drastic reduction in pH coming out of the desired alkalinity level in the system and this leads to the methanogens death. The literature reports that the optimal pH range for methanogens growth is between 6.5 and 7.5 [12]. This means that, in order to obtain an efficient production of biomethane, it is necessary to use a high organic loading rate until this feed causes a reduction of pH below a value of 6.5. Consistently with those stated previously, this happens for an OLR equal to 6 [$kg_{VS}/m^3/day$]. At this point, considering the aspects described formerly, it would be interesting to get a correlation that allows to describe the variations of pH level of the system as a function of the OLR fed to the reactor. So, through experimental data derived from different papers, it has been collected information about pH and the corresponding OLR values obtaining the following correlation (Fig. 5.8).

The considered model, a second order polynomial, can fit very well the experimental points with a R^2 equal to 0.75:

$$pH = -0.1683 \cdot OLR^2 + 1.5303 \cdot OLR + 3.7355 \qquad (5.9)$$

Focusing now on the physical meaning of the graph above, it's correct to say that the represented behaviors are coherent with those expected; in fact, increasing OLR, it's possible to reach the optimal pH range for methanogens growth, particularly for an OLR between 2.5 and 5.5 [$kg_{VS}/m^3/day$]. On the other hand, getting an OLR equal to 6 [$kg_{VS}/m^3/day$], it can be observed that the pH tends to decrease until causing

Fig. 5.8 Correlation between pH and OLR. Data have been gathered from literature [10–16]

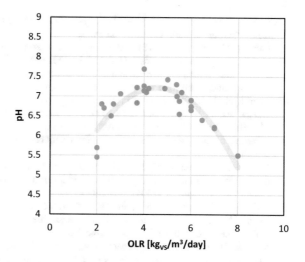

the achievement of an acid environment for high OLR like 8 [$kg_{VS}/m^3/day$]. So, in conclusion, it's important to point out two aspects related to the previous correlation:

- It can be considered suitable to describe the variation of alkalinity level inside the reactor by changing the total amount of VS fed to the anaerobic digester;
- It cannot be considered right for all the case study; in fact, as said at the beginning, OLR is a parameter that depends on various factors, so it is possible that the process inhibition occurs for OLR lower or higher than 6 [$kg_{VS}/m^3/day$].

References

1. Lili M, Biró G, Sulyok E, Petis M, Borbély J, Tamás J (2011) Novel approach on the basis of FOS/TAC method. Analele Univ Din Oradea Fasc Protecţia Mediu 17:713–718
2. Pontoni L, Panico A, Salzano E, Frunzo L, Iodice P, Pirozzi F (2015) Innovative parameters to control the efficiency of anaerobic digestion process. Chem Eng Trans 43:2089–2094. https://doi.org/10.3303/CET1543349
3. Liu X, André L, Mercier-Huat M, Grosmaître J-M, Pauss A, Ribeiro T (2021) Accurate estimation of bicarbonate and acetic acid concentrations with wider ranges in anaerobic media using classical FOS/TAC titration method. Appl Sci 11(24):24. https://doi.org/10.3390/app112411843
4. Veluchamy C, Gilroyed BH, Kalamdhad AS (2019) Process performance and biogas production optimizing of mesophilic plug flow anaerobic digestion of corn silage. Fuel 253:1097–1103. https://doi.org/10.1016/j.fuel.2019.05.104
5. Steffen R, Szolar O, Braun R (1998) Feedstocks for anaerobic digestion. Inst Agrobiotechnol Tulin Univ Agric Sci Vienna 12:1–29
6. Legner M, McMillen DR, Cvitkovitch DG (2019) Role of dilution rate and nutrient availability in the formation of microbial biofilms. Front Microbiol 10:916. https://doi.org/10.3389/fmicb.2019.00916

7. Babaee A, Shayegan J (2020) Effect of organic loading rates (OLR) on production of methane from anaerobic digestion of vegetables waste
8. Labatut RA, Pronto JL (2018) Sustainable waste-to-energy technologies: anaerobic digestion
9. Gou C, Yang Z, Huang J, Wang H, Xu H, Wang L (2014) Effects of temperature and organic loading rate on the performance and microbial community of anaerobic co-digestion of waste activated sludge and food waste
10. Chen L et al (2022) Effects of organic loading rates on the anaerobic co-digestion of fresh vinegar residue and pig manure: focus on the performance and microbial communities
11. Li L, He Q, Ma Y, Wang X, Peng X (2015) Dynamics of microbial community in a mesophilic anaerobic digester treating food waste: relationship between community structure and process stability
12. Yu Q, Feng L, Zhen X (2021) Effects of organic loading rate and temperature fluctuation on the microbial community and performance of anaerobic digestion of food waste
13. Demirel B, Yenigun O (2004) Anaerobic acidogenesis of dairy wastewater: the effects of variations in hydraulic retention time with no pH control
14. Khan MA et al (2016) Optimization of process parameters for production of volatile fatty acid, biohydrogen and methane from anaerobic digestion
15. Varol A, Ugurlu A (2017) Comparative evaluation of biogas production from dairy manure and co-digestion with maize silage by CSTR and new anaerobic hybrid reactor
16. Nagao N et al (2012) Maximum organic loading rate for the single-stage wet hanaerobic digestion of food waste

Appendix

Here is a link to the complete database, as well as its averaged variants PAD and SAD: https://tinyurl.com/28ox9lck. If you have any difficulties, please contact the authors. To contribute with new data, please send an email with your material to the authors. As a result, the database will be updated soon. Welcome. You can use the data in the database as you like. If you used the database for your work, cite our referenced article (as stated in the license in the readme file). To utilize the database, send the request from google drive with a motivation and the databases will be unlocked soon for you. The database organized as follow:

1. Complete Database: a detailed version of the database that includes all the substrate's attributes, with multiple entries for each substrate. Both proximate and ultimate analytical data are available. It is suggested to use it for data comparison, statistical analysis, and machine learning.
2. Primary Averaged Database (PAD): All the data for each substrate has been linearly averaged to obtain a single, reliable, value that can identify and characterize the substrates' properties uniquely. It is suggested to use it for process modeling and simulations.
3. Secondary Averaged Database (SAD): PAD data is averaged again, resulting in a single value for each category of every attribute in the database (Manure, Agricultural Waste, Sludge, Organic Waste). It is suggested to use it for preliminary analysis and pre-feasibility studies.

Bibliography

1. Ramos I, Fdz-Polanco F (2013) The potential of oxygen to improve the stability of anaerobic reactors during unbalanced conditions: results from a pilot-scale digester treating sewage sludge. Bioresour Technol 140:80–85. https://doi.org/10.1016/j.biortech.2013.04.066
2. Ramos I, Pérez R, Reinoso M, Torio R, Fdz-Polanco M (2014) Microaerobic digestion of sewage sludge on an industrial-pilot scale: the efficiency of biogas desulphurisation under different configurations and the impact of O_2 on the

© The Editor(s) (if applicable) and The Author(s), under exclusive license
to Springer Nature Switzerland AG 2024
F. Moretta and G. Bozzano, *Mathematical and Statistical Approaches for Anaerobic Digestion Feedstock Optimization*, SpringerBriefs in Energy,
https://doi.org/10.1007/978-3-031-56460-4

microbial communities. Bioresour Technol 164:338–346. https://doi.org/10.1016/j.biortech.2014.04.109

3. Choi I, Lin R, Shin Y (2023) Canonical correlation-based model selection for the multilevel factors. J Econ 233(1):22–44. https://doi.org/10.1016/j.jeconom.2021.09.008

4. Zhong W, Zhang T, Zhu Y, Liu JS (2012) Correlation pursuit: forward stepwise variable selection for index models. J R Stat Soc Ser B Stat Methodol 74(5):849–870. https://doi.org/10.1111/j.1467-9868.2011.01026.x

5. Shieh GS (1998) A weighted Kendall's tau statistic. Stat Probab Lett 39(1):17–24. https://doi.org/10.1016/S0167-7152(98)00006-6

6. Samara B, Randles RH (1988) A test for correlation based on Kendall's tau. Commun Stat Theory Methods 17(9):3191–3205. https://doi.org/10.1080/03610928808829798

7. Bayard R, Liu X, Benbelkacem H, Buffière P, Gourdon R (2016) Can biomethane potential (BMP) be predicted from other variables such as biochemical composition in lignocellulosic biomass and related organic residues? BioEnergy Res 9:610–623. https://doi.org/10.1007/s12155-015-9701-3

8. Xin Y, Wang D, Li XQ, Yuan Q, Cao H (2018) Influence of moisture content on cattle manure char properties and its potential for hydrogen rich gas production. J Anal Appl Pyrolysis 130:224–232. https://doi.org/10.1016/j.jaap.2018.01.005

9. Aguilar-Aguilar FA, Longoria A et al (2019) Optimization of hydrogen yield from the anaerobic digestion of crude glycerol and swine manure. Catalysts 9(4):4. https://doi.org/10.3390/catal9040316

10. Sidhu F, Safar KM, Memon M (2019) Assessment of yard waste for biological treatment in the Mehran UET, Jamshoro, 2119. https://doi.org/10.1063/1.5115380

11. Safarian S, Unnthorsson R, Richter C (2020) Simulation and Performance Analysis of Integrated Gasification–Syngas Fermentation Plant for Lignocellulosic Ethanol Production. Fermentation 6:68. https://doi.org/10.3390/fermentation6030068

12. Yao J, Wang Z, Liu M, Bai B, Zhang C (2023) Nitrate-Nitrogen Adsorption Characteristics and Mechanisms of Various Garden Waste Biochars. Mater Basel Switz 16:5726. https://doi.org/10.3390/ma16165726

13. Vassilev SV, Vassileva CG, Song Y-C, Li W-Y, Feng J (2017) Ash contents and ash-forming elements of biomass and their significance for solid biofuel combustion. Fuel 208:377–409. https://doi.org/10.1016/j.fuel.2017.07.036

14. Tosti L, van Zomeren A, Pels JR, Comans RNJ (2021) Evaluating biomass ash properties as influenced by feedstock and thermal conversion technology towards cement clinker production with a lower carbon footprint. Waste Biomass Valoriz 12(8):4703–4719. https://doi.org/10.1007/s12649-020-01339-0

15. Yang G et al (2018) Free ammonia-based sludge treatment reduces sludge production in the wastewater treatment process. Chemosphere 205:484–492. https://doi.org/10.1016/j.chemosphere.2018.04.140

16. Kato T, Kojima K, Sumikura M, Kuroiwa Y (2023) Study on ammonia generation from digested sludge by subcritical water treatment. Water Sci Technol 87(3):614–619. https://doi.org/10.2166/wst.2023.015

17. Grasham O, Dupont V, Cockerill T, Camargo-Valero MA (2022) Ammonia and biogas from anaerobic and sewage digestion for novel heat, power and transport applications—a techno-economic and GHG emissions study for the United Kingdom. Energies 15(6):6. https://doi.org/10.3390/en15062174

18. Ebner JH, Labatut RA, Lodge JS, Williamson AA, Trabold TA (2016) Anaerobic co-digestion of commercial food waste and dairy manure: characterizing biochemical parameters and synergistic effects. Waste Manag 52:286–294. https://doi.org/10.1016/j.wasman.2016.03.046

19. Nielfa A, Cano R, Fdz-Polanco M (2015) Theoretical methane production generated by the co-digestion of organic fraction municipal solid waste and biological sludge. Biotechnol Rep 5:14–21. https://doi.org/10.1016/j.btre.2014.10.005

20. Neves L, Oliveira R, Alves MM (2006) Anaerobic co-digestion of coffee waste and sewage sludge. Waste Manag 26(2):176–181. https://doi.org/10.1016/j.wasman.2004.12.022

21. Poulsen TG, Adelard L (2016) Improving biogas quality and methane yield via co-digestion of agricultural and urban biomass wastes. Waste Manag 54:118–125. https://doi.org/10.1016/j.wasman.2016.05.020

22. Xie T, Xie S, Sivakumar M, Nghiem LD (2017) Relationship between the synergistic/antagonistic effect of anaerobic co-digestion and organic loading. Int Biodeterior Biodegrad 124:155–161. https://doi.org/10.1016/j.ibiod.2017.03.025

23. Montesdeoca-Pichucho NB, Garibaldi-Alcívar K, Baquerizo-Crespo RJ, Gómez-Salcedo Y, Pérez-Ones O, Pereda-Reyes I (2023) Synergistic and antagonistic effects in anaerobic co-digestion: analysis of the methane yield kinetics. Rev Fac Ing Univ Antioquia 107:107. https://doi.org/10.17533/udea.redin.20220473

24. Batstone DJ et al (2002) The IWA Anaerobic Digestion Model No 1 (ADM1). Water Sci Technol 45(10):65–73. https://doi.org/10.2166/wst.2002.0292

25. Moretta F, Rizzo E, Manenti F, Bozzano G (2021) Enhancement of anaerobic digestion digital twin through aerobic simulation and kinetic optimization for co-digestion scenarios. Bioresour Technol 341:125845. https://doi.org/10.1016/j.biortech.2021.125845

Printed in the United States
by Baker & Taylor Publisher Services